Coastal Construction Manual

Principles and Practices of Planning, Siting, Designing, Constructing, and Maintaining Residential Buildings in Coastal Areas (Fourth Edition)

FEMA P-55 / Volume I / August 2011

FEMA

Preface

The 2011 *Coastal Construction Manual*, Fourth Edition (FEMA P-55), is a two-volume publication that provides a comprehensive approach to planning, siting, designing, constructing, and maintaining homes in the coastal environment. Volume I provides information about hazard identification, siting decisions, regulatory requirements, economic implications, and risk management. The primary audience for Volume I is design professionals, officials, and those involved in the decision-making process.

Volume II contains in-depth descriptions of design, construction, and maintenance practices that, when followed, will increase the durability of residential buildings in the harsh coastal environment and reduce economic losses associated with coastal natural disasters. The primary audience for Volume II is the design professional who is familiar with building codes and standards and has a basic understanding of engineering principles.

Acknowledgments

Fourth Edition Authors and Key Contributors

William Coulbourne, Applied Technology Council
Christopher P. Jones, Durham, NC
Omar Kapur, URS Group, Inc.
Vasso Koumoudis, URS Group, Inc.
Philip Line, URS Group, Inc.
David K. Low, DK Low and Associates
Glenn Overcash, URS Group, Inc.
Samantha Passman, URS Group, Inc.
Adam Reeder, Atkins
Laura Seitz, URS Group, Inc.
Thomas Smith, TLSmith Consulting
Scott Tezak, URS Group, Inc. – Consultant Project Manager

Fourth Edition Volume I Reviewers and Contributors

Marcus Barnes, FEMA Headquarters
Mark Crowell, FEMA Headquarters
Lois Forster, FEMA Headquarters
John Ingargiola, FEMA Headquarters – Technical Assistance and Research Contracts Program Manager
Tucker Mahoney, FEMA Headquarters
Alan Springett, FEMA Region II
Paul Tertell, FEMA Headquarters – Project Manager
Mark Vieira, FEMA Region IV
Jonathan Westcott, FEMA Headquarters
David Zaika, FEMA Headquarters
Dana Bres, U.S. Department of Housing and Urban Development
Stuart Davis, U.S. Army Corps of Engineers
Roy Domangue, Wooden Creations, Inc.
Brad Douglas, American Forest and Paper Association
Russell J. Coco, Jr., Engensus
Carol Friedland, Louisiana State University
Trudie Johnson, Town of Hilton Head Island
Ernie Katzwinkel, Dewberry
Vladimir Kochkin, National Association of Home Builders
Stephen Leatherman, Florida International University
Amit Mahadevia, URS Group, Inc.
Peter Mazikins, American Forest and Paper Association
Deborah Mills, Dewberry
Manuel Perotin, Atkins
Rebecca Quinn, RCQuinn Consulting, Inc.
Billy Ward, Champion Builders, LLC

Fourth Edition Technical Editing, Layout, and Illustration

Diana Burke, URS Group, Inc.
Lee-Ann Lyons, URS Group, Inc.
Susan Ide Patton, URS Group, Inc.
Billy Ruppert, URS Group, Inc.

Contents

List of Figures

Chapter 3

Chapter 4

Chapter 5

Chapter 6

List of Tables

1

Introduction

1.1   Background

The Federal Emergency Management Agency (FEMA) first published the *Coastal Construction Manual* (FEMA 55) in 1981. The Manual was updated in 1986 and provided guidance to public officials, designers, architects, engineers, and contractors for over a decade. In that time, however, construction practices and materials changed, and more information on hazards and building performance was developed and used to update the Manual again in 2000.

Over the past several decades, the coastal population in the United States has increased significantly. The increased coastal population led to increased coastal development, which led in turn to greater numbers of structures at risk from coastal hazards. Additionally, many of the residential buildings constructed today are larger and more valuable than those of the past, resulting in the potential for larger economic losses when disasters strike. A FEMA study estimates that the combination of population growth and sea level rise may increase the portion of the U.S. population residing in a coastal floodplain from 3 percent in 2010 to as much as 4 percent in 2100 (FEMA 2010a [draft]).

In response to increased hazards and lessons learned from past storms, regulatory requirements for construction in coastal areas have increased over the past decade. In 2000, the International Code Council (ICC) created the International Code Series (I-Codes) based on the three regional model building codes: the Building Officials Code Administrators

CROSS REFERENCE

Regulatory requirements, including the I-Codes, CZMA, and the NFIP, are addressed in Chapter 5.

The *Coastal High Hazard Area* (or Zone V) is explained in Section 3.6.2 of this Manual.

International (BOCA) National Building Code (NBC), the Southern Building Code Congress International (SBCCI) Southern Building Code (SBC), and the International Conference of Building Officials (ICBO) Uniform Building Code (UBC). Based on data included in the Insurance Services Office (ISO) Building Code Effectiveness Grading Schedule (BCEGS) database, 86.5 percent of jurisdictions in the hurricane-prone region have adopted wind-resistant building codes, and 47.25 percent of flood-prone jurisdictions have adopted flood-resistant building codes (ISO 2011). As of the publication of this Manual, 33 of the 35 coastal States and U.S. territories, in implementing the Coastal Zone Management Act (CZMA) of 1972, have instituted construction setbacks and coastal resource protection programs. Many jurisdictions now require geotechnical studies and certifications from design professionals for construction along the coastline. Finally, as of May 2011, over 21,450 communities participate in the National Flood Insurance Program (NFIP), which requires, among other things, that plans for new buildings constructed in Coastal High Hazard Areas be certified by a design professional.

Investigations conducted by FEMA and other organizations after major coastal disasters have consistently shown that properly sited, well-designed, and well-constructed coastal residential buildings generally perform well (refer to Chapter 2 for a discussion of the FEMA investigations). This updated *Coastal Construction Manual*—prepared by FEMA with assistance from other agencies, organizations, and professionals involved in coastal construction and regulation—is intended to help designers and contractors identify and evaluate practices that will improve the quality of construction in coastal areas and reduce the economic losses associated with coastal disasters.

The design and construction techniques included in this Manual are based on a comprehensive evaluation of:

- Coastal residential buildings, both existing and under construction

- Siting, design, and construction practices employed along the U.S. coastlines

- Building codes, floodplain management ordinances, and standards applicable to coastal construction

- Performance of coastal buildings based on post-disaster field investigations

1.2   Purpose

This Manual provides guidance for designing and constructing residential buildings in coastal areas that will be more resistant to the damaging effects of natural hazards. The focus is on new residential construction and substantial improvement or repairs of substantial damage to existing residential buildings—principally detached single-family homes, attached single-family homes (townhouses), and low-rise (three-story or less) multi-family buildings. Some of the recommendations of the Manual may also apply to non-substantial improvements or repairs. Discussions, examples, and example problems are provided for buildings in or near coastal flood hazard areas in a variety of coastal environments subject to high winds, flooding, seismic activity, erosion, and other hazards.

This Manual is intended to be used by contractors, designers, architects, and engineers who are familiar with the design and construction of one- to three-story residential buildings in coastal areas of the United States and its territories. Readers less familiar with design and construction practices, as well as State and community officials, should also refer to FEMA P-762, *Local Officials Guide for Coastal Construction* (FEMA 2009), for guidance on planning and design considerations for improving the performance of coastal residential buildings before using this Manual.

1.3   Objectives

The goal of this Manual is to provide professionals guidance to assist them in pre-design, planning tasks and decisions as well as design and construction practices that will lead to building successful, disaster-resistant homes. For any project, it is critical that the project be well planned in order to minimize potential issues later on during the design and construction process and when the building is impacted by an event. These items are summarized in the following sections and elaborated on in detail throughout this Manual.

1.3.1   Planning for Construction

One objective of this Manual is to highlight the many tasks and decisions that must be made ***before actual construction begins***. These tasks include, but may not be limited to:

- Evaluating the suitability of coastal lands for residential construction

- Planning for development of raw land and for infill or redevelopment of previously developed land

- Identifying regulatory, environmental, and other constraints on construction or development

- Evaluating site-specific hazards and loads at a building site

- Evaluating techniques to mitigate hazards and reduce loads

- Identifying risk, insurance, and financial implications of siting, design, and construction decisions

1.3.2   Successful Buildings

A second objective of this Manual is to identify the best design and construction practices for ***building successful disaster-resistant structures***.

In coastal areas, a building can be considered successful only if it is capable of resisting damage from coastal hazards and processes over a period of decades. This does not mean that a coastal residential building will remain undamaged over its intended lifetime, but that undermining from erosion and the effects of a design-level flood or wind event (or series of lesser events with combined impacts equivalent to a design event) will be limited.

> **NOTE**
>
> The designer should be familiar with the recommendations in this Manual, along with the building codes and engineering standards cited, as these may establish an expected level of professional care.

A ***successful building*** is considered a building for which the following are true after a design-level event:

- The building foundation is intact and functional

- The envelope (lowest floor, walls, openings, and roof) is structurally sound and capable of minimizing penetration of wind, rain, and debris

- The lowest floor elevation is high enough to prevent floodwaters from entering the building envelope

- The utility connections (e.g., electricity, water, sewer, natural gas) remain intact or can be easily restored

- The building is accessible and habitable

- Any damage to enclosures below the lowest floor does not result in damage to the foundation, utility connections, or elevated portions of the building or nearby structures

- For buildings affected by a design level seismic event, the building protects life and provides safety, even if the structure itself sustains significant damage

1.3.2.1 Premise and Framework for Achieving Successful Designs

The underlying goal of a successful design is expressed through its basic premise: ***Anticipated loads must be transferred through the building in a continuous path to the supporting soils.*** Any weakness in that continuous path is a potential point of failure. To fulfill this design premise, designers must address a variety of issues and constraints. These are illustrated in Figure 1-1 and summarized as follows:

Funding. Any project is constrained by available funding, and designers must balance building size and expense against the

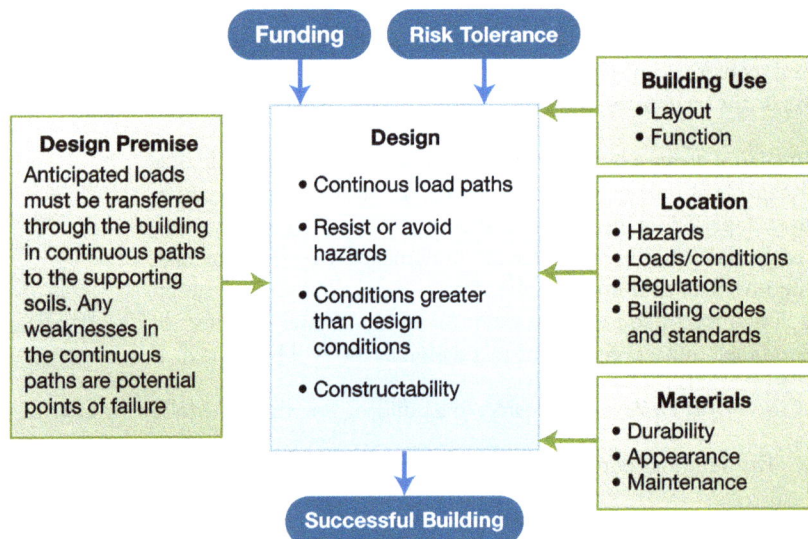

Figure 1-1. Design framework to achieve successful buildings

desire for building success. Initial and long-term costs should be factored into the design. Higher initial construction costs may result in increased closing costs or higher mortgage rates, but may minimize potential building damage, reduce insurance rates, and reduce future maintenance costs.

Risk tolerance. Some owners are willing and able to assume a high degree of financial and other risks, while other owners are more conservative and seek to minimize potential building damage and future costs.

Building use. The intended use of the building will affect its layout, form, and function.

Location. The location of the building will determine the nature and intensity of hazards to which the building will be exposed; loads and conditions that the building must withstand; and building codes, standards, and regulations that must be satisfied.

Materials. A variety of building materials are available, and some are better suited to coastal environments than others. Owners and designers must select appropriate materials that address both aesthetic and durability issues. If an owner is prepared for frequent maintenance and replacement, the range of available materials will be wider; however, most owners are not prepared to do so, and the most durable materials should be used.

Continuous load paths. Continuous load paths must be constructed and maintained over the intended life of the building.

Resist or avoid hazards. The magnitudes of design forces acting on structures, coupled with project funding, building location, and other factors, will determine which forces can be resisted and which must be avoided. Structures are typically designed to resist wind loads and avoid flood loads (through elevation on strong foundations).

Conditions greater than design conditions. Design loads and conditions are based on some probability of exceedance, and it is always possible that design loads and conditions can be exceeded. Designers can anticipate this and modify their initial design to better accommodate higher forces and more extreme conditions. The benefits of doing so often exceed the costs of building higher and stronger.

Constructability. Ultimately, designs will only be successful if they can be implemented by contractors. Complex designs with many custom details may be difficult to construct and could lead to a variety of problems, both during construction and once the building is occupied.

1.3.2.2 Best Practices Approach

To promote best practices, portions of the Manual recommend and advocate techniques that exceed the minimum requirements of model building codes; design and construction standards; or Federal, State, and local regulations. The authors of the Manual are aware of the implications of such recommendations on the design, construction, and cost of coastal buildings, and make them only after careful review of building practices and subsequent building performance during design level events.

Some of the recommended best practices and technical solutions presented in the previous version of FEMA 55 (2000, third edition) have been incorporated into the model building codes. For example:

■ The 2009 and 2012 editions of the International Residential Code (IRC)—see sections R322.2.1(2) and R322.3.2(1)—require 1 foot of ***freeboard*** in the Coastal A Zone and in certain Zone V situations. Past minimum code provisions did not require any freeboard. Note that more than 1 foot of freeboard may be indicated once the design framework steps outlined in Figure 1-1 are accomplished.

- The 2006, 2009, and 2012 editions of the International Building Code (IBC) require conformance with American Society of Civil Engineers (ASCE) Standard 24-05, *Flood Resistant Design and Construction*. ASCE 24-05 requires new buildings situated in the Coastal A Zone to be designed and constructed to Zone V requirements. Thus, the 2000 version of the *Coastal Construction Manual* recommendation to treat Coastal A Zone

TERMINOLOGY: FREEBOARD

Freeboard is an additional height that buildings are elevated above the base flood elevation (BFE). Freeboard acts as a factor of safety to compensate for uncertainties in the determination of flood elevations, and provides an increased level of flood protection. Freeboard will result in reduced flood insurance premiums.

buildings like Zone V buildings is now being implemented for IBC-governed buildings through the building code.

Sustainable building design concepts are increasingly being incorporated into residential building design and construction through green building rating systems. While the environmental benefits associated with adopting green building practices can be significant, these practices must be implemented in a manner that does not compromise the building's resistance to natural hazards. FEMA P-798, *Natural Hazards and Sustainability for Residential Buildings* (FEMA 2010b), examines current green building rating systems in a broader context. It identifies green building practices—the tools of today's green building rating systems—that are different from historical residential building practices and that, unless implemented with an understanding of their interactions with the rest of the structure, have the potential to compromise a building's resistance to natural hazards. FEMA P-798 discusses how to retain or improve natural hazard resistance while incorporating green building practices.

1.4 Organization and Use of This Manual

This Manual first provides a history of coastal disasters in the United States, an overview of the U.S. coastal environment, and fundamental considerations for constructing a building in a coastal region. The Manual covers every step in the process of constructing a home in a coastal area: evaluating potential sites; selecting a site; locating, designing, and constructing the building; and insuring and maintaining the building. Flowcharts, checklists, maps, equations, and details are provided throughout the Manual to help the reader understand the entire process. In addition, example problems are presented to demonstrate decisions and calculations designers must make to reduce the potential for damage to the building from natural hazard events.

The Manual also includes numerous examples of siting, design, and construction practices—both good and bad—to illustrate the results and ramifications of those practices. The intent is twofold: (1) to highlight the benefits of practices that have been employed successfully by communities, designers, and contractors, and (2) to warn against practices that have resulted in otherwise avoidable damage or loss of coastal residential buildings.

1.4.1   Organization

Because of its size, the Manual is divided into two volumes, with a total of 15 chapters. Additional supporting materials and resources are available at the FEMA Residential Coastal Construction Web site.

Volume I

Chapter 1 – Introduction. This chapter describes the purpose of the Manual, outlines the content and organization, and explains how icons are used throughout the Manual to guide and advise the reader.

Chapter 2 – Historical Perspective. This chapter summarizes selected past coastal flood and wind events and post-event evaluations, and other major milestones. It documents the causes and types of damage associated with storms and tsunamis ranging from the 1900 hurricane that struck Galveston, TX, to the Samoan tsunami that struck American Samoa following an earthquake in September 2009.

Chapter 3 – Identifying Hazards. This chapter describes coastal processes, coastal geomorphology, and coastal hazards. Regional variations for the Great Lakes, North Atlantic, Middle Atlantic, South Atlantic, Gulf of Mexico, Pacific, Alaska, Hawaii, and U.S. territories are discussed. This chapter also discusses hazards that influence the design and construction of a coastal building (coastal storms, erosion, tsunamis, and earthquakes) and their effects.

Chapter 4 – Siting. This chapter describes the factors that should be considered when selecting building sites, including small parcels in areas already developed, large parcels of undeveloped land, and redevelopment sites. Guidance is also provided to help designers and contractors determine how a building should be placed on a site. Detailed discussions of the coastal construction process begin in this chapter.

Chapter 5 – Investigating Regulatory Requirements. This chapter presents an overview of building codes and Federal, State, and local regulations that may affect construction on a coastal building site. Additionally, the NFIP, Coastal Barrier Resources Act (CBRA), and Coastal Zone Management (CZM) programs are described.

Chapter 6 – Fundamentals of Risk Analysis and Risk Reduction. This chapter summarizes acceptable levels of risk; tradeoffs in decisions concerning siting, design, construction, and maintenance; and cost and insurance implications that should be considered in coastal construction.

Volume II

Chapter 7 – Pre-Design Considerations. This chapter introduces the design process, minimum design requirements, inspections, and sustainable design considerations. It discusses the cost and insurance implications of decisions made during design and construction. It also outlines the contents of Volume II.

Chapter 8 – Determining Site-Specific Loads. This chapter explains how to calculate site-specific loads, including loads from high winds, flooding, seismic events, and tsunamis, as well as combinations of more than one load. Example problems are provided to illustrate the application of design load provisions of ASCE 7-10, *Minimum Design Loads for Buildings and Other Structures* (ASCE 2010).

Chapter 9 – Designing the Building. This chapter contains information on designing each part of a building to withstand expected loads. Topics covered include structural failure modes, load paths, building systems, application of loads, structural connections, building material considerations, requirements for breakaway walls, and considerations for designing appurtenances.

Chapter 10 – Designing the Foundation. This chapter presents recommendations for the selection and design of foundations. Design of foundation elements including pile capacity in soil, installation methods, and material durability considerations are discussed.

Chapter 11 – Designing the Building Envelope. This chapter describes how to design roof coverings, exterior wall coverings, exterior doors and windows, shutters, and soffits to resist natural hazards.

Chapter 12 – Mechanical Equipment and Utilities. This chapter provides guidance on design considerations of mechanical equipment and utilities, as well as techniques that can improve the capability of equipment to survive a natural disaster.

Chapter 13 – Constructing the Building. This chapter describes how to properly construct a building in a coastal area and how to avoid common construction mistakes that may lessen the ability of a building to withstand a natural disaster. It includes guidance on material choices and durability, and construction techniques for improved resistance to decay and corrosion.

Chapter 14 – Maintaining the Building. This chapter explains special maintenance concerns for new and existing buildings in coastal areas. Methods to reduce damage from corrosion, moisture, weathering, and termites are discussed, along with building elements that require frequent maintenance.

Chapter 15 – Retrofitting Existing Buildings. This chapter includes broad guidance for evaluating existing residential structures to assess the need and feasibility for wildfire, seismic, flood, and wind retrofitting. It also includes a discussion of wind retrofit packages that encourage homeowners to take advantage of opportunities to strengthen their homes while performing routine maintenance (e.g., roof shingle replacement).

Resources and Supporting Material

The FEMA Residential Coastal Construction Web site (http://www.FEMA.gov/rebuild/mat/fema55.shtm) provides guidance and other information to augment the content of this Manual. The material provided on the Web site includes a glossary for this Manual as well as:

- **Resource documents.** Examples include *Dune Walkover Guidance, Material Durability in Coastal Environments,* and *Swimming Pool Design Guidance.*

- **Links and contact information.** Government agencies, professional and trade organizations, code and standard organizations, and natural hazard and coastal science organizations.

- **Links to additional Web sites and coastal construction resources published by FEMA.** Examples include the *Wind Retrofit Guide for Residential Buildings* (FEMA P-804), *Home Builder's Guide to Coastal Construction* (FEMA P-499), and the FEMA Safe Room and Building Science Web sites.

NOTE

In previous editions of the Coastal Construction Manual, Volume III contained appendices and information that expanded on content provided in Volumes I and II. The FEMA Residential Coastal Construction Web site now serves as the location for additional content.

1.4.2   Using the Manual

This Manual uses icons as visual guides to help readers quickly find information. These icons call out notes, warnings, definitions, cross references, cost considerations, equations, example problems, and specific hazards.

Notes. Notes contain supplemental information that readers may find helpful, including things to consider when undertaking a coastal construction project, suggestions that can expedite the project, and the titles and sources of other publications related to coastal construction. Full references for publications are presented at the end of each chapter of the Manual.

Warnings. Warnings present critical information that will help readers avoid mistakes that could result in dangerous conditions, violations of ordinances or laws, and possibly delays and higher costs in a coastal construction project. Any questions about the meanings of warnings in this Manual should be directed to the appropriate State or local officials.

Terminology. The meanings of selected technical and other special terms are presented where appropriate.

Cross references. Cross references point the reader to information that supplements or further explains issues of interest in this Manual, such as technical discussions, regulatory information, equations, tables, and figures.

Cost Considerations. Cost consideration notes discuss issues that can affect short-term and lifecycle and insurance costs associated with a coastal residential construction project.

Equations. Volume II includes equations for calculating loads imposed by forces associated with natural hazard events. It also presents equations used in the design of building components intended to withstand the loads imposed by design events. Equations are numbered for ease of reference.

Examples. In Volume II, example problems demonstrate the calculation of flood, wind, and seismic loads on a coastal residential building. Example problems are numbered for ease of reference.

1.4.3   Hazard Icons

Hazard icons will help readers find information specific to their needs (see below). To use the icons effectively, readers must determine in which flood zone the property or building site in question is located. Chapter 3 of this Manual explains how to make such a determination and includes detailed definitions of the flood hazard zones.

Zone V. Portion of the *Special Flood Hazard Area* (SFHA) that extends from offshore to the inland limit of a primary frontal dune along an open coast, and any other area subject to high-velocity wave action from storms or tsunamis.

Coastal A Zone. A subset of Zone A. Specifically, that portion of the SFHA landward of Zone V (or landward of a coastline without a mapped Zone V) in which the principal source of flooding is coastal storms, and where the potential base flood wave height is between 1.5 and 3.0 feet.

Zone A. Portion of the SFHA in which the principal source of flooding is runoff from rainfall, snowmelt, or coastal storms where the potential base flood wave height is between 0.0 and 3.0 feet.

> **TERMINOLOGY: SPECIAL FLOOD HAZARD AREA**
>
> The SFHA is the land area covered by the floodwaters of the base flood on NFIP maps. It is the area where the NFIP's floodplain management regulations must be enforced and the area where the mandatory purchase of flood insurance applies. The SFHA includes Zones A, AO, AH, A1-30, AE, A99, AR, AR/A1-30, AR/AE, AR/AO, AR/AH, AR/A, VO, V1-30, VE, and V.

Zone X. Includes shaded and unshaded Zone X. The flood hazard is less severe here than in the SFHA.

1.4.4. Contact Information

Every effort has been made to make this Manual as comprehensive as possible. However, no single manual can anticipate every situation or need that may arise in a coastal construction project. Readers who have questions not addressed herein should consult local officials. Information is also available from the FEMA Building Science Helpline (Web: http://www.fema.gov/rebuild/buildingscience/, e-mail: FEMA-Buildingsciencehelp@dhs.gov, telephone: 866-927-2104), and the Mitigation Division of the appropriate FEMA Regional Office. Contact information for FEMA personnel, the State NFIP Coordinating Agencies, and the State Coastal Zone Management Agencies are provided on the FEMA Residential Coastal Construction Web page.

1.5   References

ASCE (American Society of Civil Engineers). 2005. *Flood Resistant Design and Construction*. ASCE Standard ASCE 24-05.

ASCE. 2010. *Minimum Design Loads for Buildings and Other Structures*. ASCE Standard ASCE 7-10.

FEMA (Federal Emergency Management Agency). 2000. FEMA 55 (3rd Edition). *Coastal Construction Manual: Principles and Practices of Planning, Siting, Designing, Constructing, and Maintaining Residential Buildings in Coastal Areas*. May.

FEMA. 2009. P-762, *Local Officials Guide for Coastal Construction: Design Considerations, Regulatory Guidance, and Best Practices for Coastal Communities*. February.

FEMA. 2010a (draft). *The Impact of Climate Change on the National Flood Insurance Program*. August 2010 draft; still draft as of July 2011).

FEMA. 2010b. P-798, *Natural Hazards and Sustainability for Residential Buildings*. September.

ISO (Insurance Service Office, Inc.). 2011. *Building Code Effective Grading Schedule*. http://www.isomitigation.com/bcegs/0000/bcegs0001.html. Accessed June 28, 2011.

Historical Perspective

2.1   Introduction

Through the years, FEMA, other Federal agencies, State and local agencies, and other private groups have documented and evaluated the effects of coastal flood and wind events and the performance of buildings located in coastal areas during those events. These evaluations provide a historical perspective on the siting, design, and construction of buildings along the Atlantic, Pacific, Gulf of Mexico, and Great Lakes coasts. These studies provide a baseline against which the effects of later coastal flood events can be measured.

Within this context, certain hurricanes, coastal storms, and other coastal flood events stand out as being especially important, either because of the nature and extent of the damage they caused or because of particular flaws they exposed in hazard identification, siting, design, construction, or maintenance practices. Many of these events—particularly those occurring since 1979—have been documented by FEMA in Flood Damage Assessment Reports, Building Performance Assessment Team (BPAT) reports, and Mitigation Assessment Team (MAT) reports. These reports summarize investigations that FEMA conducts shortly after major disasters. Drawing on the combined resources of a Federal, State, local, and private sector partnership, a team of investigators

CROSS REFERENCE

For resources that augment the guidance and other information in this Manual, see the Residential Coastal Construction Web site (http://www.fema.gov/rebuild/mat/fema55.shtm).

NOTE

Hurricane categories reported in this Manual should be interpreted cautiously. Storm categorization based on wind speed may differ from that based on barometric pressure or storm surge. Also, storm effects vary geographically—only the area near the point of landfall will experience effects associated with the reported storm category.

is tasked with evaluating the performance of buildings and related infrastructure in response to the effects of natural and man-made hazards. The teams conduct field investigations at disaster sites; work closely with local and State officials to develop recommendations for improvements in building design and construction; and prepare recommendations concerning code development, code enforcement, and mitigation activities that will lead to greater resistance to hazard events.

This chapter summarizes coastal flood and wind events that have affected the United States and its territories since the beginning of the twentieth century. The lessons learned regarding factors that contribute to flood and wind damage are discussed.

2.2   Coastal Flood and Wind Events

This section summarizes major coastal flood and wind events in the United States from 1900 to 2010. Many of these events have led to changes in building codes, regulations, mapping, and mitigation practices. The map and timeline in Figure 2-1 provide a chronological list of the major coastal flood and wind events in combination with the major milestones resulting from the events. They show the evolution of coastal hazard

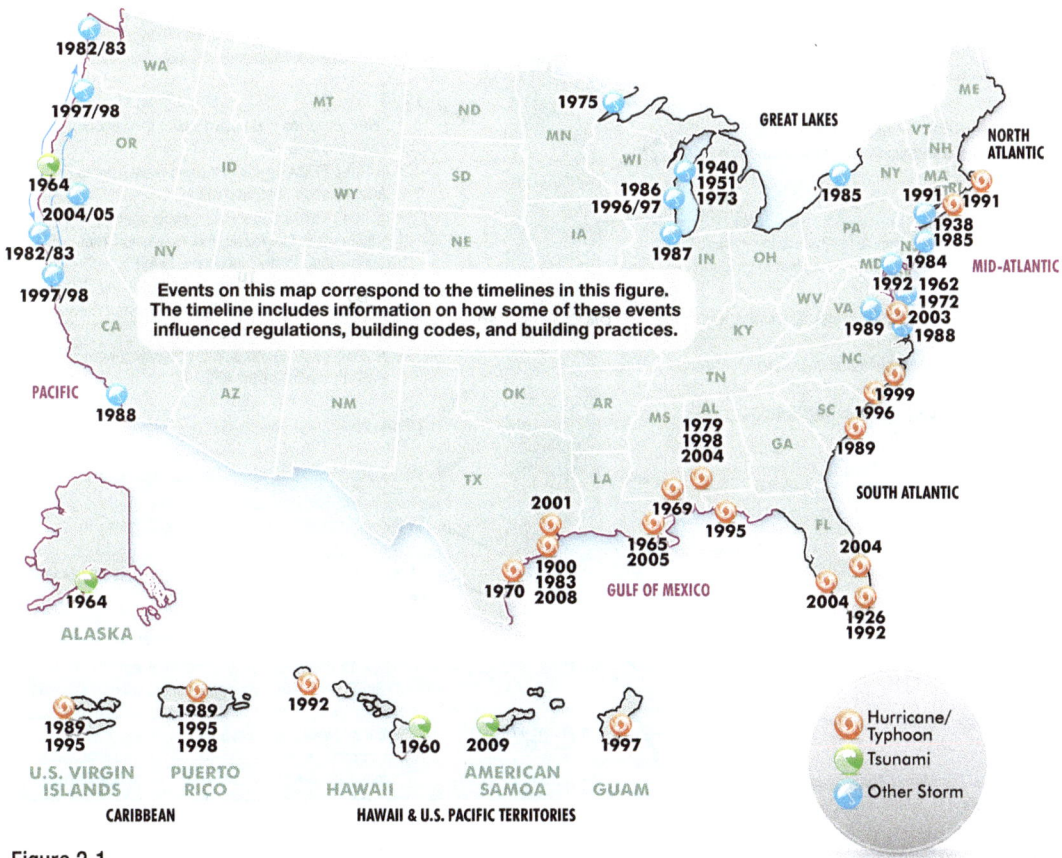

Figure 2-1.
Map and timeline of significant coastal flood and wind events, and milestones for regulations, building codes, and building practices

1900 GALVESTON HURRICANE September, Galveston, TX
- City of Galveston initiates a large-scale project to raise ground elevations and buildings.

1938 NEW ENGLAND HURRICANE September, Long Island, NY/New England

1960 TSUNAMI April, Hilo, HI

1964 GOOD FRIDAY EARTHQUAKE & TSUNAMI March, AK/CA

1968 Congress establishes the NFIP

1969 HURRICANE CAMILLE August, MS/AL
- In 1971, National Bureau of Standards post-storm report concludes, "...damage directly attributable to wave action and flooding far exceeded that due to wind... Greater consideration should be given to storm surge..."

1972 TROPICAL STORM AGNES June, Mid-Atlantic.
- Flood Disaster Protection Act of 1973 includes Mandatory Flood Insurance Purchase Requirement

1975 GREAT LAKES STORM November, Western Great Lakes

1981 NFIP establishes methodology to assess contribution of wave runup to BFEs; the methodology is applied in ME.

1982/83 WINTER COASTAL STORMS CA/OR/WA
- In 1985, conference concludes that siting standards are needed for building in areas subject to erosion.

1983 HURRICANE ALICIA September, Galveston, and Houston, TX
- TDI is formed to develop formal inspection process for wind damage.
- One of the first post-hurricane buyout programs begins in Baytown, TX.

1985 HURRICANE GLORIA September, NY/NJ
- NJ implements new coastal development practices.

1986 FEMA publishes second edition of FEMA 55

1986 GREAT LAKES STORM WI

1988 WINTER COASTAL STORM January, Southern CA

1988 First edition of ASCE 7 published

1989 NOR'EASTER March, Nags Head and Kill Devil Hills, NC/Sandbridge Beach, VA

1900
1910
1920
1930
1940
1950
1960
1970
1980
1990

(continued on page 2-5)

1926 MIAMI HURRICANE September, Miami, FL
- In 1927, Local engineer's post-storm inspection report stresses the importance of proper design, construction quality, and implementation of building codes.

1940 ARMISTICE DAY STORM November, Lake Michigan

1951 STORM November, Lake Michigan

1962 NOR'EASTER March, Mid-Atlantic

1965 HURRICANE BETSY September, FL/LA
- Flooding from storm leads to a major redesign of the levee system by the USACE.
- Congress passes Southeast Hurricane Disaster Relief Act mandating a study of disaster insurance options.
- In 1968, Congress passes the National Flood Insurance Act, which creates the NFIP.

1970 HURRICANE CELIA August, Corpus Christi, TX
- In 1971, Texas Catastrophe Property Insurance Association (TCPIA) and Texas Wind Insurance Association (TWIA) are formed [precursor to Texas Department of Insurance (TDI)].

1973 NOR'EASTER April, Lake Michigan
- NFIP requires elevation to the 100-year flood.

1979 HURRICANE FREDERIC September, AL
- FEMA performs first post-disaster investigation after Hurricane Frederic.
- In 1980, FEMA begins to include wave heights in determination of coastal BFEs.
- In 1980, Mobile County, AL, adopts specific requirements for glazing, roof overhangs, roof reinforcements, and anchoring. In 1985, these measures performed well during Hurricane Elena.
- In 1981, FEMA publishes the first edition of FEMA 55, *Coastal Construction Manual.*
- In 1983, FEMA recommends breakaway walls on grade level enclosures below BFE.

1984 NOR'EASTER March, NJ

1985 GREAT LAKES STORMS March, Great Lakes

1987 GREAT LAKES STORM February, Chicago, IL

1988 NOR'EASTER April, Sandbridge Beach, VA/Nags Head, NC

1989 HURRICANE HUGO September, SC/PR
- FEMA's first building performance assessment team (BPAT) documents poor performance of roof systems, which later, after Hurricane Andrew, leads to changed roof and wall sheathing attachment practices and awareness of continuous load paths. These observations also lead to roof test methods and standards.
- FEMA BPAT recommends Coastal A Zone, sufficient pile embedment, and enforcement of building code wind design requirements.

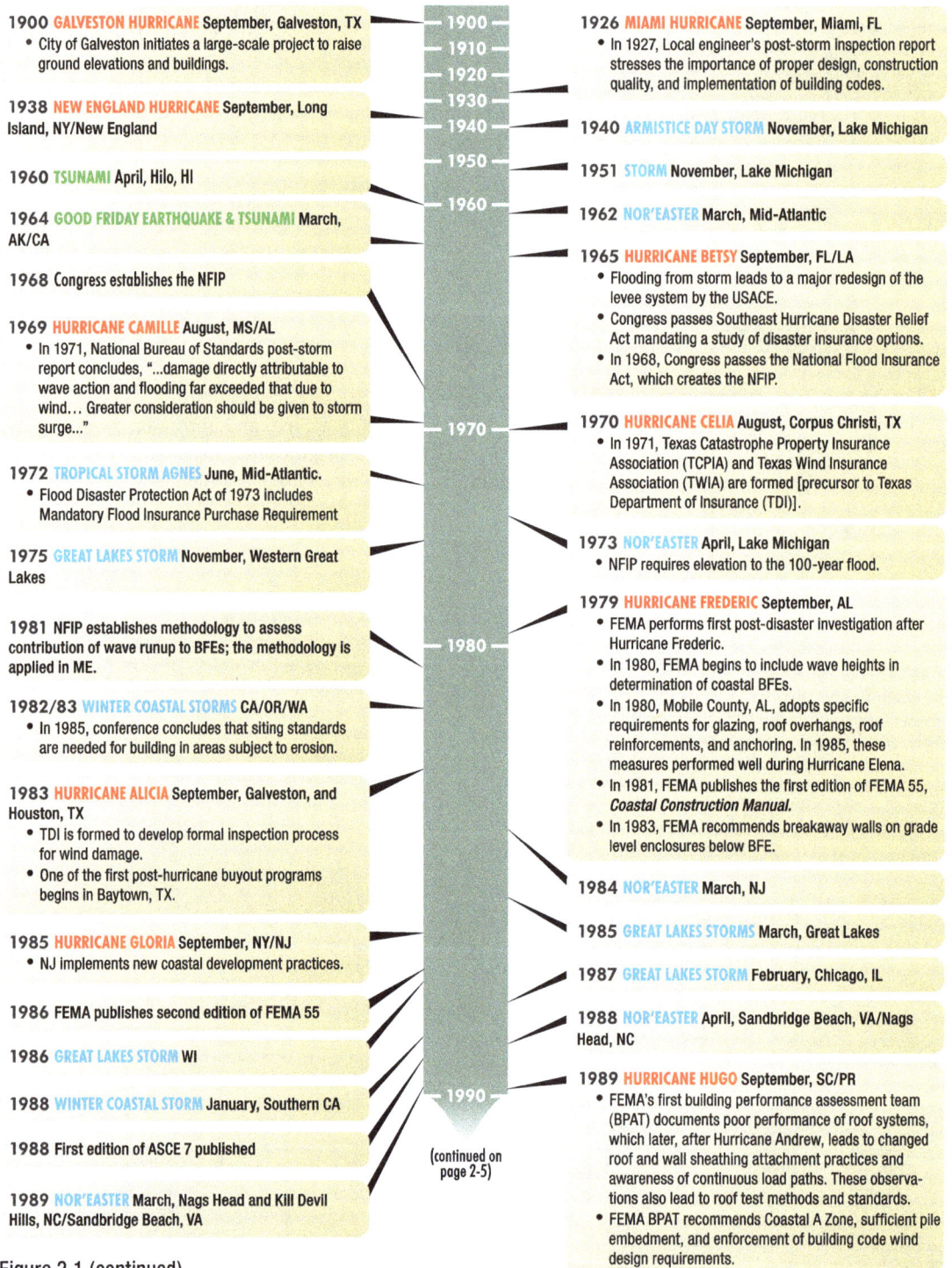

Figure 2-1 (continued).
Map and timeline of significant coastal flood and wind events, and milestones for regulations, building codes, and building practices

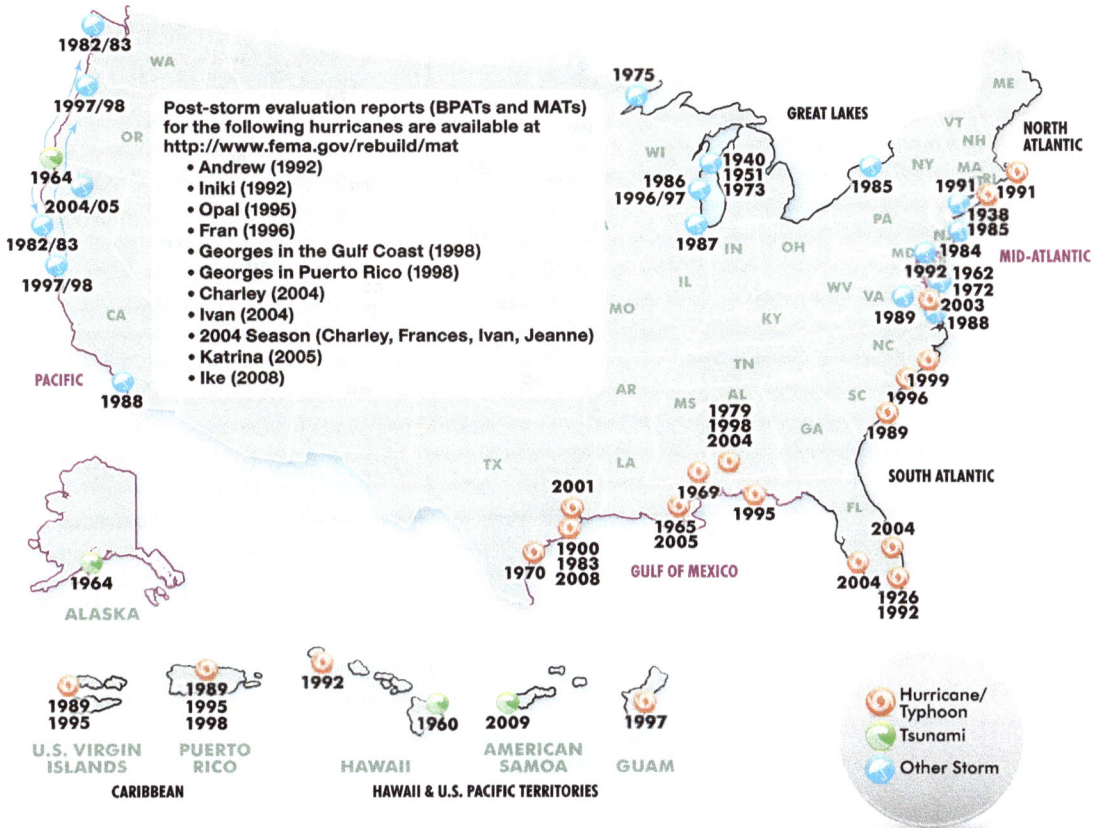

Post-storm evaluation reports (BPATs and MATs) for the following hurricanes are available at http://www.fema.gov/rebuild/mat
- Andrew (1992)
- Iniki (1992)
- Opal (1995)
- Fran (1996)
- Georges in the Gulf Coast (1998)
- Georges in Puerto Rico (1998)
- Charley (2004)
- Ivan (2004)
- 2004 Season (Charley, Frances, Ivan, Jeanne)
- Katrina (2005)
- Ike (2008)

Figure 2-1 (continued).
Map and timeline of significant coastal flood and wind events, and milestones for regulations, building codes, and building practices

mitigation practices in the United States since the year 1900. Each event is color-coded by hazard type and corresponds to a symbol on the map where the storm occurred. The map shows the eight coastal regions defined in this chapter.

2.2.1   North Atlantic Coast

The North Atlantic Coast is generally considered the coastal area from northern Maine to Long Island, NY. This coastal area is most susceptible to nor'easters and hurricane remnants, but significant hurricanes occasionally make landfall. Flood and erosion damage is often significant, damaging foundations and even undermining buildings to the point of collapse. Wind causes roof and envelope damage, especially as a result of tree fall.

CROSS REFERENCE

For a more detailed history of storms for the different areas of the United States see the Residential Coastal Construction Web site (http://www.fema.gov/rebuild/mat/fema55.shtm).

The National Oceanic and Atmospheric Administration (NOAA) provides detailed tropical storm and hurricane track information starting in 1848 (http://csc.noaa.gov/hurricanes/).

(continued from page 2-3)

— 1990 —

1990 NFIP Community Rating System (CRS) begins implementation.

1991 HURRICANE BOB August, Buzzards Bay Area, MA

1992 NOR'EASTER January, DE/MD

1991 NOR'EASTER October, Long Island, NY/Eastern MA

1992 HURRICANE INIKI September, Kauai County, HI
- BFEs for Hawaii are recalculated to include hurricane flood effects in addition to tsunami effects.

1992 HURRICANE ANDREW August, FL
- Existing State wind pools gain momentum.
- Designers recognize the vulnerability and importance of the building envelope.
- APA produces guidance for roof sheathing attachment.
- Dade and Broward County governments are the first to enact provisions for windborne debris (1993 SFBC and 1995 ASCE 7)
- In 1994, HUD adopts more stringent wind design criteria for manufactured homes. These measures performed well during Hurricane Georges in 1998 and Hurricane Charley in 2004.

1995 HURRICANE MARILYN September, USVI
- USVI adopts a current model code, replacing the outdated code.

1995 HURRICANE OPAL October, FL Panhandle

1995 FEMA publishes first edition of FEMA 259 *Engineering Principles & Practices for Retrofitting Flood Prone Residential Buildings.*

1996 HURRICANE FRAN September, Southeastern NC
- FEMA BPAT reiterates need for Coastal A Zone.

1996/97 GREAT LAKES WINTER STORMS WI

1997/98 WINTER COASTAL STORM Pacific Coast

1997 TYPHOON PAKA December, Guam
- Guam adopts ASCE 7, which includes the influence of topography in wind speed.

1998 First edition of ASCE 24 published

1998 HURRICANE GEORGES September, PR/MS/AL/FL
- Puerto Rico adopts a current model code.

1999 HURRICANE FLOYD September, Mid-Atlantic
- North Carolina takes State ownership of its mapping program.
- Along with other hurricanes, reveals issues with flood insurance in CBRA zones.

2000 FEMA publishes third edition of FEMA 55.

2000 ICC publishes the first International Code Series.

— 2000 —

2001 TROPICAL STORM ALLISON June, Houston,TX

2004 HURRICANE FRANCES September, FL

2003 HURRICANE ISABEL September, Mid-Atlantic

2004 HURRICANE IVAN September, AL
- In response to extensive storm surge and flooding, FEMA begins mapping production to identify the flood damage extent. If adopted by communities, the maps will allow claims to be paid in non-SFHAs. This is the forerunner to the post-Katrina ABFE mapping.

2004 HURRICANE CHARLEY August, FL
- IBHS begins developing FORTIFIED program to build and retrofit safer residential buildings.

2004/05 SEVERE WINTER STORMS CA

2004 HURRICANE JEANNE September, FL

2005 FEMA publishes first edition of FEMA 499, *Homebuilder's Guide to Coastal Construction Fact Sheet Series.*

2005 HURRICANE KATRINA September, LA/MS
- Mississippi and Louisiana adopt current model codes. Previous codes were outdated or non-existent.
- FEMA begins release of advisory BFEs and recovery maps for the post-Katrina Gulf Coast. Communities are encouraged to adopt the ABFE maps to guide redevelopment until complete restudy of the flood risk is complete.
- In 2006, FEMA develops pre-engineered coastal foundations and publishes FEMA 550, *Recommended Residential Construction for Coastal Areas.*
- In 2007, FEMA publishes FEMA 543, *Design Guide for Improving Critical Facility Safety from Flooding and High Winds.*

2008 FEMA Procedure Memorandum 50 establishes guidelines for mapping the Limit of Moderate Wave Action (LiMWA).

2008 HURRICANE IKE September, Galveston, TX

2009 SAMOAN TSUNAMI September, American Samoa

2009 Hawaii State Building Code adopts special wind region maps.

2009 IRC mandates freeboard in Zone V and Coastal A Zone.

2009 FEMA publishes FEMA P-762, *Local Officials Guide to Coastal Construction.*

— 2010 —

2010 FEMA publishes FEMA P-804, *Wind Retrofit Guide for Residential Buildings.*

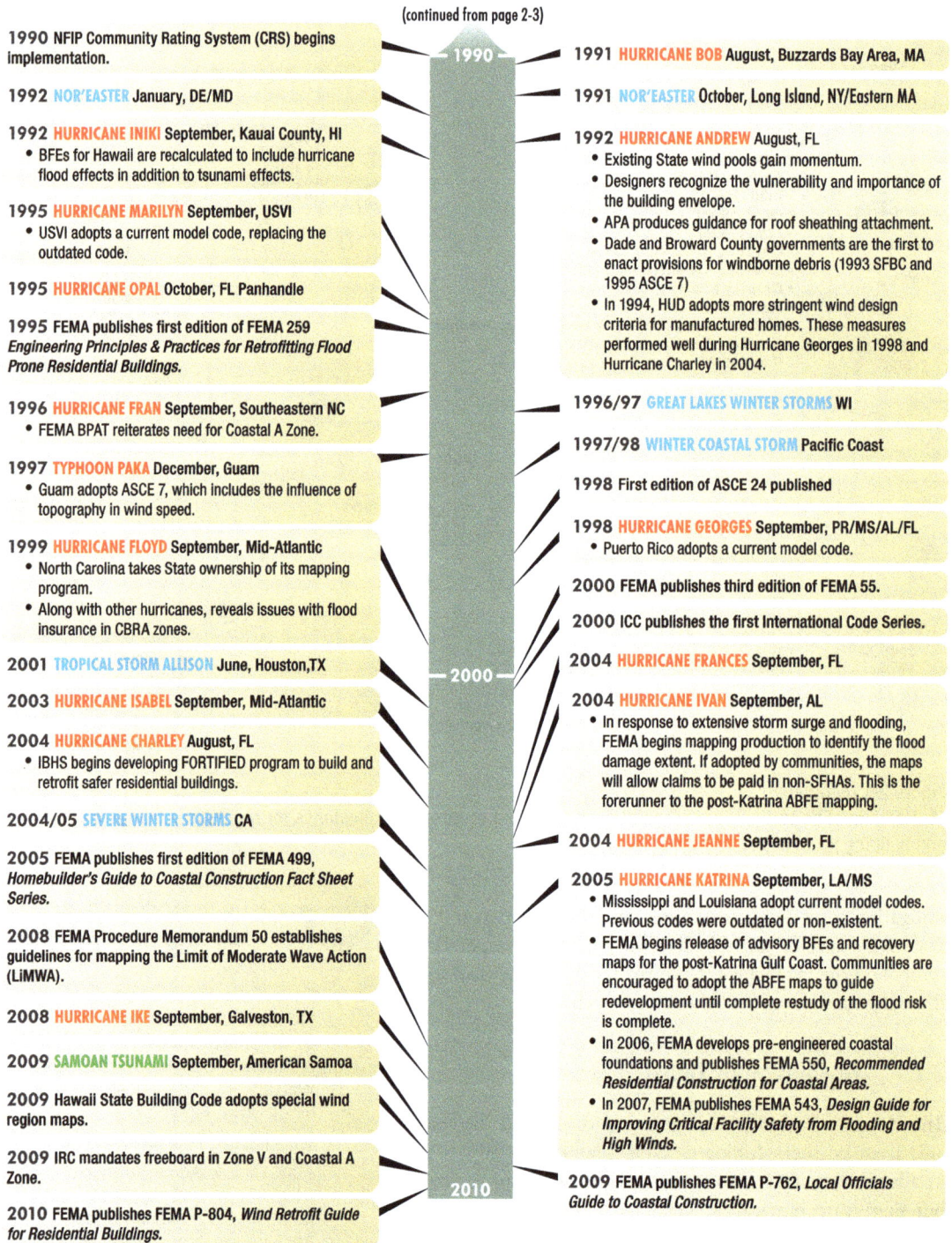

Figure 2-1 (concluded).
Map and timeline of significant coastal flood and wind events, and milestones for regulations, building codes, and building practices

In 1938, the *"Long Island Express" hurricane* moved rapidly up the east coast from New York through New England. The storm caused widespread surge and wind damage to buildings, and is still used as a benchmark for predicting worst-case scenario damage in the region (Figure 2-2). Although not shown in the photograph, this hurricane also destroyed many elevated homes along this stretch of coastline.

In September 1985, *Hurricane Gloria* hit Long Island, NY, and New Jersey, causing minor storm surge and erosion damage and significant wind damage. In 1991, New England was hit by two major storms— *Hurricane Bob* in August and a *nor'easter* in October. A FEMA Flood Damage Assessment Report noted that flood damage to buildings constructed before the local adoption of the Flood Insurance Rate Map (FIRM), known as pre-FIRM construction, that had not been elevated or that had not been elevated sufficiently suffered major damage, while properly elevated buildings constructed after the adoption of the FIRM (post-FIRM) performed well (URS 1991c). These storms provided insight into successful foundation design practices.

Figure 2-2.
Schell Beach before and after the Long Island Express Hurricane in 1938; houses near the shoreline were destroyed and more distant houses were damaged (Guilford, CT)

SOURCE: WORKS PROGRESS ADMINISTRATION PHOTOGRAPH FROM MINSINGER 1988

2.2.2   Mid-Atlantic Coast

The Mid-Atlantic Coast is generally considered the coastal area from New Jersey to Virginia. This coastal area is susceptible to both nor'easters and hurricanes with flood and wind damage similar to the damage that occur in New England.

In March 1962, a significant nor'easter, known as the *Great Atlantic Storm of 1962* or the *Ash Wednesday Storm*, affected almost the entire eastern seaboard and caused extreme damage in the Mid-Atlantic region. The combination of sustained high winds with spring tides resulted in severe beachfront erosion and flooding, sweeping many buildings out to sea.

In June 1972, *Tropical Storm Agnes* produced rains up to 19 inches, resulting in severe riverine flooding from New York to Virginia and billions of dollars in flood damage. The catastrophic damage from this storm led to the "Mandatory Flood Insurance Purchase Requirement" in the Flood Disaster Protection Act of 1973 (see Section 5.2 for more on the history of the NFIP).

A *March 1984 nor'easter* caused significant erosion problems. As a result of damage observed after this storm and *Hurricane Gloria* (see Section 2.2.1), New Jersey implemented several changes to its coastal development practices in 1985.

An *April 1988 nor'easter* caused foundation damage to elevated homes in Virginia and North Carolina. Long-term shoreline erosion, coupled with the effects of three previous coastal storms, had left the area vulnerable. Inspections following the 1988 nor'easter revealed that repairs to previous foundation damage were only partially effective. In some cases, ineffective repairs implemented after storms resulted in subsequent storm damage that may not have occurred if the original repair had been properly made (URS 1989). A *March 1989 nor'easter* in the same area caused even further foundation damage. The damage from the 1988 and 1989 storms showed that long-term erosion makes buildings increasingly vulnerable (Figure 2-3) to the effects of even minor storms (URS 1990).

A few years later, an intense *January 1992 nor'easter* hit Delaware and Maryland. Observations made by the FEMA BPAT after this storm noted damage due to storm surge, wave action, and erosion, as well as many load path failures in coastal buildings (FEMA 1992).

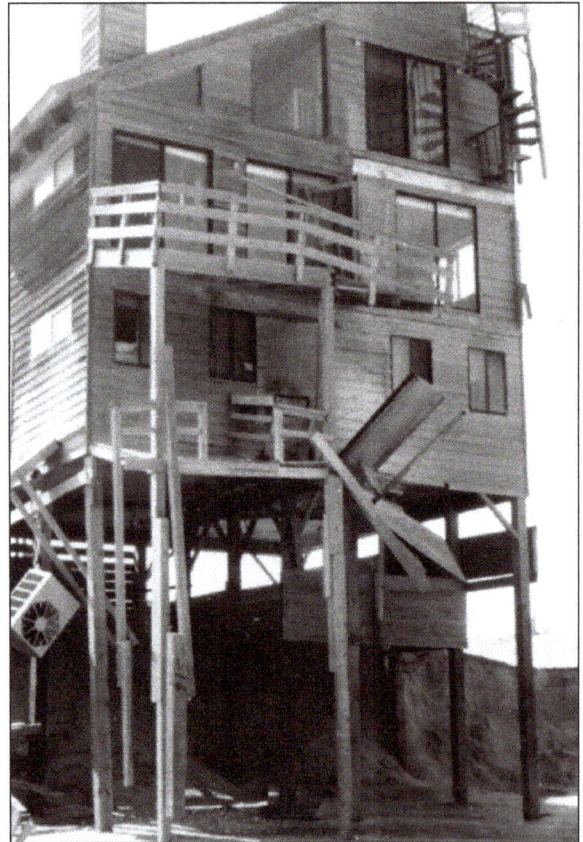

Figure 2-3.
Although this house seems to have lost only several decks and a porch during the March 1989 nor'easter, the loss of supporting soil due to long-term erosion left its structural integrity in question following successive storms

In September 2003, **_Hurricane Isabel_** made landfall near Cape Lookout, NC, as a Category 2 hurricane, breaching the barrier island. Storm surge and heavy rainfall caused extensive flooding across the Mid-Atlantic region, especially in areas adjacent to the Chesapeake Bay. Maximum observed water levels at stations along the Chesapeake Bay exceeded historical observations (NOAA 2004).

2.2.3   South Atlantic Coast

The South Atlantic Coast is generally considered the coastal area from North Carolina up to and including the Florida Keys. This region, especially the North Carolina Outer Banks and south Florida, is often subjected to hurricanes. States in the northern part of this region, such as North Carolina, are also susceptible to nor'easters. Damage is typically caused by flooding, waves, erosion, water-borne debris, wind, and wind-borne debris. The degree of damage ranges from slight to severe, depending on the characteristics of the storm.

After a **_September 1926 hurricane_** hit Miami, FL, a south Florida engineer, Theodore Eefting, wrote an article on the damage pointing out many weaknesses in buildings and construction that continue to be discussed today. Most notably, he stressed the consequences of poor quality construction, and the importance of strengthening building codes (Eefting 1927).

In late September 1989, **_Hurricane Hugo_** struck South Carolina. Observations following this hurricane revealed notable differences between the performance of pre- and post-FIRM buildings. Additionally, the BPAT deployed after Hurricane Hugo noted that some of the most severely damaged buildings were several rows back from the shoreline, and as a result recommended that design standards for Coastal A Zones (defined in Chapter 1) be more stringent. The wind damage from Hurricane Hugo also exposed deficiencies in residential roofing practices (URS 1991a, URS 1991b, and Texas Tech 1990).

In August 1992, **_Hurricane Andrew_** struck the southeast Atlantic coast. This hurricane remains one of the most memorable hurricanes to hit this region and one of the costliest to date. The majority of the damage from this hurricane was due to wind; many of the failures were traced to inadequate connections between building elements (Figure 2-4). As such, buildings could not resist wind forces because of the lack

Figure 2-4.
Roof structure failure due to inadequate bracing and inadequate fastening of the roof deck, Hurricane Andrew (Dade County, FL, 1992)

of continuous load transfer paths from the roofs to the foundations (FEMA 1993). Hurricane Andrew was a major catalyst for building code changes involving wind design that improved wind pressure calculation procedures and emphasized the need for a continuous load transfer path in buildings for uplift and lateral loads, not just for the traditional downward-acting gravity loads. Hurricane Andrew destroyed 97 percent of the manufactured homes in its path, leading the Department of Housing and Urban Development (HUD) to adopt more stringent wind design criteria for manufactured homes (FEMA 2009a).

In 1996, **Hurricane Fran** hit North Carolina. The resulting wave damage reinforced the idea that buildings in Coastal A Zones should be more hazard-resistant. The FEMA BPAT report noted that more stringent design codes and standards were needed to achieve improved performance (FEMA 1997).

In September 1999, **Hurricane Floyd** briefly touched Florida before making landfall in North Carolina and moving north along the east coast as a tropical storm all the way to Maine. Although inland flood damage was severe in eastern North Carolina, high winds, storm surge and torrential rains caused moderate damage to coastal and inland communities along much of the east coast.

2.2.4   Gulf of Mexico Coast

The Gulf of Mexico coast includes the coastal area from the Florida Keys northward and westward to Texas. This coastal area has long been susceptible to strong hurricanes, and in recent years the northern Gulf Coast (Florida panhandle to east Texas) has experienced a number of them. Low-lying areas are especially vulnerable to damage from erosion, waves, and storm surge.

The **September 1900 hurricane** that hit Galveston, TX, is still the deadliest natural disaster to affect the United States. Shortly after, as a result of destruction due to poor siting practices, Galveston Island completed the first large-scale retrofit project in the United States: roads and hundreds of buildings were elevated, ground levels in the city were raised several feet, and the Galveston seawall was built (Walden 1990). In 1961, the extensive damage caused by erosion from Hurricane Carla again highlighted the need for proper siting and construction in coastal areas (Hayes 1967).

Hurricane Camille, a Category 5 hurricane, made landfall in Mississippi in August 1969 and caused "near total destruction" in some areas near the beach as a result of waves and storm surge (Thom and Marshall 1971). High winds also caused damage farther inland. The studies performed by Thom and Marshall after the hurricane led to building design criteria that resulted in the construction of new homes with improved resistance to higher wind forces.

In September 1979, **Hurricane Frederic** hit Alabama and caused widespread damage, including the destruction of many houses elevated to the BFE. After Hurricane Frederic, FEMA began to include wave heights in its determination of BFEs in coastal flood hazard areas (FEMA 1980).

TERMINOLOGY

BASE FLOOD ELEVATION (BFE):
The BFE is the water surface elevation resulting from a flood that has a 1 percent chance of equaling or exceeding that level in any given year. Section 3.6.1 has more information on how the BFE is established.

DESIGN FLOOD ELEVATION (DFE):
The DFE is the locally adopted regulatory flood elevation. If a community regulates to minimum NFIP requirements, the DFE is identical to the BFE. If a community chooses to exceed minimum NFIP requirements, the DFE exceeds the BFE.

Hurricane Alicia made landfall in August 1983 in the Houston-Galveston area, causing extensive wind and flood damage. Wood frame houses were the hardest hit, and most of the damage was traced to poor roof construction and inadequate roof-to-wall connections (National Academy of Sciences 1984). Homes near the water were washed off their foundations, leading to the recommendation that grade-level enclosures be constructed with breakaway walls.

> **NOTE**
>
> The NFIP regulates structures to the BFE while building codes regulate to the DFE. The DFE is either equivalent to or greater than the BFE, depending on the governing codes of the jurisdiction in which the structure is located.

In October 1995, ***Hurricane Opal*** hit the Florida panhandle, exacerbating erosion and structural damage from a weaker hurricane (Hurricane Erin) that hit the area 1 month earlier. A FEMA BPAT revealed that post-FIRM Zone A and pre-FIRM buildings failed most often, especially those with insufficient pile embedment. In addition, damage observations confirmed that State regulations that exceeded NFIP requirements helped reduce storm damage (FEMA 1996).

Hurricane Georges made landfall in Mississippi in September 1998 and moved north and east through Alabama and Florida, causing both flood and wind damage. The FEMA BPAT found that buildings constructed in accordance with building codes and regulations, and buildings using specialized materials such as siding and roof shingles designed for higher wind speeds, performed well. The FEMA BPAT also confirmed that manufactured homes built after 1994 (when HUD wind design criteria were adopted following Hurricane Andrew) performed well. Most of the observed flood damage was attributed to inadequately elevated and improperly designed foundations, as well as poor siting practices (FEMA 1999a).

In June 2001, ***Tropical Storm Allison*** made landfall in Galveston, TX. It took a unique path, stalling and then making a loop around Houston, resulting in heavy rainfall of more than 30 inches over a 4-day period. Severe flooding destroyed over 2,700 homes in Houston (RMS 2001). Flood damage to commercial and government buildings in the greater Houston area was severe. Tropical Storm Allison made it clear that some of the most destructive tropical systems are not hurricanes, but slow-moving tropical storms dropping large amounts of rainfall.

Hurricane Charley made landfall in Florida in August 2004. After observing extensive wind damage, the FEMA MAT concluded that buildings built to the 2001 Florida Building Code (FBC) generally performed well structurally (FEMA 2005a), but older buildings experienced damage because design wind loads underestimated wind pressures on some building components, buildings lacked a continuous load path, and building elements were poorly constructed and poorly maintained.

In September 2004, ***Hurricane Ivan*** made landfall in Alabama and Florida. Although not a design wind event, Ivan caused extensive envelope damage that allowed heavy rains to infiltrate buildings and damage interiors. This damage highlighted weaknesses in older building stock and the need for improved guidance and design criteria for better building performance at these "below code" events. Flood-borne debris and wave damage extended into Coastal A Zones (FEMA 2005b).

In August 2005, ***Hurricane Katrina*** caused extensive storm surge damage and flooding well beyond the SFHA in Louisiana and Mississippi. Flooding in New Orleans was worsened by levee failures, and floodwaters rose well above the first floor of elevated buildings (Figure 2-5). The long duration of the flooding added to the destruction (FEMA 2006). After Katrina, FEMA issued new flood maps for the area that built on the

Figure 2-5.
This elevated house atop a masonry pier foundation was lost, probably due to waves and storm surge reaching above the top of the foundation, Hurricane Katrina (Long Beach, MS, 2005)

hazard knowledge gained in the 25+ years since the original FIRMs for that area were published. These flood maps continue to aid in rebuilding stronger and safer Gulf Coast communities.

In September 2008, **Hurricane Ike** made landfall over Galveston, TX, and although wind speeds were below design levels, storm surge was more characteristic of a Category 4 hurricane. High waves and storm surge destroyed or substantially damaged over two-thirds of the buildings on Bolivar Peninsula. The FEMA MAT recommended enforcement of the Coastal A Zone building requirements that were recommended in earlier editions of the Coastal Construction Manual and discussed in Chapter 5 of this Manual, as well as designing critical facilities to standards that exceed current codes (FEMA 2009b).

2.2.5   U.S. Caribbean Territories

The U.S. Caribbean Territories of the U.S. Virgin Islands and Puerto Rico are frequently hit by tropical storms and hurricanes. Damage in the Caribbean Territories is generally made worse by poor construction practices and less stringent building codes.

In 1989, **Hurricane Hugo** destroyed many buildings in the U.S. Virgin Islands and Puerto Rico (York 1989). In 1995, the U.S. Virgin Islands and Puerto Rico were again struck by a hurricane. High winds from **Hurricane Marilyn** damaged roofs (Figure 2-6), allowing water to penetrate and damage building interiors (National Roofing Contractors Association [NRCA] 1996). This storm highlighted the need for more stringent building codes, and the U.S. Virgin Islands adopted the 1994 UBC.

In 1998, the high winds and flooding from **Hurricane Georges** caused extensive structural damage in Puerto Rico. While not all of the damage could have been prevented, a significant amount could have been avoided if more buildings had been constructed to meet the requirements of the Puerto Rico building code and floodplain management regulations in effect at the time (FEMA 1999b). In 1999, as a result of FEMA BPAT recommendations, Puerto Rico adopted the 1997 UBC.

Figure 2-6.
This house lost most of its metal roof covering due to high winds during Hurricane Marilyn in 1995 (location unknown)
SOURCE: NRCA 1996

2.2.6 Great Lakes Coast

The Great Lakes Coast extends westward from New York to Minnesota. The biggest threat to coastal properties in the Great Lakes region is wave damage and erosion brought on by high winds associated with storms passing across the region during periods of high lake levels. Sometimes, stalled storm systems bring extremely heavy precipitation to local coastal areas, resulting in massive property damage from flooding, bluff and ravine slope erosion from storm runoff, and bluff destabilization from elevated groundwater.

> **NOTE**
>
> Lake levels in the Great Lakes fluctuate seasonally by 1 to 2 feet. High lake levels can intensify flood damage.

In November 1940, the ***Armistice Day Storm*** brought high winds and heavy rain to the eastern shoreline of Lake Michigan, tearing roofs off buildings and blowing out windows. The wind damage also uprooted trees and downed telephone and power lines.

A ***November 1951 storm*** hit Lake Michigan exacerbating already near-record high lake levels and causing extensive erosion and flooding that broke through seawalls. Damage observed as a result of this storm was consistent with the concept of Great Lakes shoreline erosion as a slow, cumulative process, driven by lakebed erosion, high water levels, and storms.

An ***April 1973 storm*** caused storm surge resulting in erosion damage around Lake Michigan. The storm caused flooding 4 feet deep in downtown Green Bay, WI. The floodwaters here reached the elevation of the 0.2-percent-annual-chance flood due to strong winds blowing along the length of the bay piling up a storm surge on already high lake levels.

A ***November 1975 storm*** hit the western Great Lakes, undermining harbors, destroying jetties, and sinking an ore carrier with its crew onboard. The storm severely undermined the harbor breakwater at Bayfield, WI, requiring its replacement the following year.

High winds from a *March 1985 storm* caused storm surge flooding in upstate New York and Lake Erie, where lake levels rose to record levels. That month, Wisconsin's Lake Michigan lakeshore suffered rapid shoreline recession in successive storms, and some homes had to be relocated.

The southeastern Wisconsin coast of Lake Michigan experienced rainfall in excess of the 0.2-percent-annual-chance precipitation event as a result of a *1986 storm*, causing massive property damage from flooding, erosion, and bluff destabilization (U.S. Army Corp of Engineers [USACE] 1997, 1998).

A *February 1987 storm* hit Chicago, IL, during a period of record high lake levels on Lake Michigan (Figure 2-7 shows damage from a similar storm). High waves destroyed a seawall and caused severe erosion to Chicago's lakeshore. Waves slammed high-rise condominiums, smashing first floor windows, and flooding basements.

The southeastern Wisconsin coast of Lake Michigan experienced *two rainfall events, in 1996 and 1997*, each of which resulted in precipitation in excess of the 0.2-percent-annual-chance event. These events, similar to the 1986 storm, caused massive property damage from flooding, erosion, and bluff destabilization (USACE 1997, 1998).

Figure 2-7.
Erosion along the Lake Michigan shoreline at Holland, MI, resulting from high lake levels and storm activity (August 1988)
SOURCE: MARK CROWELL, FEMA

2.2.7 Pacific Coast

The Pacific Coast extends from Alaska to southern California. The Pacific Coast is mostly affected by high waves and erosion during winter storms, though tsunamis occasionally affect the area. Hurricanes can affect the southern Pacific Coast, but this is rare. Damage to homes from El Niño-driven storms over the past several decades reinforces the importance of improving siting practices near coastal bluffs and cliffs on the Pacific Coast.

A *March 1964 earthquake* with an epicenter in Prince William Sound, Alaska, generated a *tsunami* that affected parts of Washington, Oregon, California, and Hawaii. The tsunami flooded entire towns and triggered landslides. A post-disaster report provided several recommendations on foundation design, such as deep foundations to resist scour and undermining, and placement of wood frame buildings (Wilson and Tørum 1968).

In the winter of 1982-83, a series of ***El Niño-driven coastal storms*** caused widespread and significant damage to beaches, cliffs, and buildings along the coast between Baja California and Washington. These storms prompted a conference on coastal erosion, which concluded that siting standards were needed for homes built in areas subject to erosion, especially those atop coastal bluffs (McGrath 1985). The California Coastal Commission now uses the 1982-83 storms as its design event for new development (California Coastal Commission, 1997).

In January 1988, a rapidly developing ***coastal storm*** struck southern California. The waves from the storm were the highest on record at the time and severely damaged shore protection structures and oceanfront buildings. This storm demonstrated the severity of damage that could be caused by a winter storm.

In the winter of 1997-98, another notable series of severe ***El Niño-driven coastal storms*** battered the coasts of California and Oregon. Heavy rainfall caused widespread soil saturation, resulting in debris flow, landslides, and bluff collapse.

California experienced ***severe storms*** in the winter of 2004-05, where heavy rain, debris flow, and landslides damaged buildings. A single landslide in Conchita, CA, destroyed 13 houses and severely damaged 23 houses in 2005 (Figure 2-8) (Jibson 2005).

Figure 2-8.
This building experienced structural damage due to a landslide in La Conchita, CA, after a January 2005 storm event
SOURCE: JOHN SHEA, FEMA

2.2.8   Hawaii and U.S. Pacific Territories

Hawaii and the U.S. Pacific Territories of Guam, the Northern Marianas Islands, and American Samoa are subject to tropical cyclones (called hurricanes in Hawaii and American Samoa, and typhoons in Guam and the Northern Marianas Islands) and tsunamis. Tropical cyclones can cause damage in these areas from high winds, large waves, erosion, and rapid flow of rainfall runoff down steep terrain. Tsunamis can cause damage from rapidly moving water and debris across the shoreline area.

In 1992, **Hurricane Iniki**, the strongest hurricane to affect the Hawaiian Islands in recent memory, caused significant flood and wave damage to buildings near the shoreline. Following the hurricane, FEMA recalculated BFEs to include hurricane flood effects, instead of just tsunami effects. This revision made flood maps more accurate and aided in the rebuilding process. A FEMA BPAT after the hurricane revealed problems with foundation construction that resulted in some buildings being washed off their foundations. It also concluded that inadequately designed roofs and generally poor quality of construction resulted in wind damage that could have been avoided.

In December 1997, **Typhoon Paka** hit Guam causing substantial damage to wood-frame buildings, but minimal damage to concrete and masonry buildings. After the typhoon, Guam adopted ASCE 7-98 design wind speeds, which incorporated topographic influences in wind speeds for the first time.

In September 2009, an 8.0 magnitude earthquake occurred approximately 160 miles southwest of American Samoa. Within 20 minutes, a series of **tsunami** waves struck the island. Due to high waves and runup, at least 275 residences were destroyed and several hundred others were damaged (Figure 2-9). Damage to commercial buildings, churches, schools and other buildings was also widespread. Elevated buildings and buildings farther inland generally performed better because they were able to avoid dynamic flood loads.

Figure 2-9.
Tsunami damage at Poloa, American Samoa
SOURCE: ASCE, USED WITH PERMISSION

2.3 Breaking the Disaster-Rebuild-Disaster Cycle

Although the physiographic features vary throughout the coastal areas of the United States, post-event damage assessments and reports show that the nature and extent of damage caused by coastal flood events are remarkably similar. Similar findings have been noted for coastal storms in which high winds damage the built environment. In the case of wind, the evolution of building for "wind resistance" is characterized by improved performance of some building components (e.g., structural systems), but continued poor performance of other elements (e.g., building envelope components).

Although many aspects of coastal design and construction have improved over the years, the harsh coastal environment continues to highlight deficiencies in the design and construction process. The design and construction community should incorporate the lessons learned from past events in order to avoid repeating past mistakes, and to break the disaster-rebuild-disaster cycle.

The conclusions of post-event assessments can be classified according to those factors that contribute to both building damage and successful building performance: hazard identification, siting, design, construction, and maintenance. Special attention must also be paid when designing and constructing enclosures in coastal buildings. Reduction of building damage in coastal areas requires attention to these factors and coordination between owners, designers, builders, and local officials.

2.3.1 Hazard Identification

Understanding and identifying the hazards that affect coastal areas is a key factor in successful mitigation. Historical and recent hurricanes have provided insight into coastal hazards and their effects on coastal buildings. An **all-hazards approach** to design is needed to address all possible impacts of coastal storms and other coastal hazards.

NOTE

Conclusions presented in this section are based on numerous post-event damage assessments by FEMA and other technical and scientific organizations. Although most of the findings are qualitative, they serve as a valuable source of information on building performance and coastal development practices.

CROSS REFERENCE

Chapter 3 discusses coastal hazards in more detail and their effects on coastal buildings.

Sections 1.4.3 and 3.3 of this Manual explain the concept of the Coastal A Zone.

WARNING

FIRMs do not account for future effects of sea level rise and long-term erosion. All mapped flood hazard zones (V, A, and X) in areas subject to sea level rise and/or long-term erosion likely underestimate the extent and magnitude of actual flood hazards that a coastal building will experience over its lifetime. FIRMs also do not account for storm-induced erosion that has occurred after the FIRM effective date.

Refer to Section 3.5 for more detailed information on erosion.

The minimum Zone A foundation and elevation requirements should not be assumed to provide buildings with resistance to coastal flood forces. The **Coastal A Zone** recommendations in this Manual should be considered as a part of the best practices approach to designing a successful building. Flood hazards in areas mapped as Zone A on coastal FIRMs can be much greater than flood hazards in riverine Zone A for two reasons:

1. Waves 1.5 to 3 feet high (i.e., too small for an area to be classified as Zone V, but still capable of causing structural damage and erosion) occur during base flood conditions in many areas.

2. Older FIRMs may fail to reflect changing site conditions (e.g., as a result of long-term erosion, loss of dunes during previous storms) and improved flood hazard mapping procedures.

Addressing **all potential flood hazards** will help reduce the likelihood of building damage or loss. The building in Figure 2-10 was approximately 1.3 miles from the Gulf of Mexico shoreline, but was damaged by storm surge and small waves during Hurricane Ike. Flood damage can result from the effects of short- and long-term increases in water levels (storm surge, tsunami, riverine flooding, poor drainage, seiche, and sea-level rise), wave action, high-velocity flows, erosion, and debris.

Failure to consider long-term hazards, such as **long-term erosion** and the **effects of multiple storms**, can increase coastal flood hazards over time. Long-term erosion and accumulation of short-term erosion impacts over time can cause loss of protective beaches, dunes, and bluffs, and soils supporting building foundations. Failure to account for long-term erosion is one of the more common errors made by those siting and designing coastal residential buildings. Similarly, failure to consider the effects of multiple storms or flood events may lead to underestimating flood hazards in coastal areas. Coastal buildings left intact by one storm may be vulnerable to damage or destruction by successive storms.

In coastal bluff areas, consideration of the potential effects of surface and subsurface drainage, removal of vegetation, and site development activities can help reduce the likelihood of **slope stability hazards and landslides**. Drainage from **septic systems** on coastal land can destabilize coastal bluffs and banks, accelerate erosion, and increase the risk of damage and loss to coastal buildings. Vertical cracks in the soils of some cohesive bluffs can cause a **rapid rise of groundwater levels** in the bluffs during extremely heavy and prolonged precipitation events. The presence of these cracks can rapidly reduce the stability of such bluffs.

High winds can cause both structural and building envelope damage. Exposure and topography can increase wind pressures and wind damage. Homes on **barrier islands and facing large bays** or bodies of water

Figure 2-10.
School located approximately 1.3 miles from the Gulf shoreline damaged by storm surge and small waves, Hurricane Ike (Cameron Parish, LA, 2008)

may be exposed to wind pressures higher than in areas of flat terrain, especially at high pressure zones of the roof. The house in Figure 2-11 sustained damage at the roof edge and roof corners, even though the hurricane was below the design event and wind damage should not have occurred. Recent studies have influenced wind design standards to increase design wind pressures on these exposed structures. Failure to consider the ***effects of topography*** (and changes in topography such as bluff erosion) on wind speeds can lead to an underestimation of design wind speeds. ***Siting buildings on bluffs*** or near high-relief topography requires special attention by the designer.

> **CROSS REFERENCE**
>
> Section 8.7.1 explains the increased wind pressures on certain zones of a roof (Figure 8-17).

Some coastal areas are also susceptible to ***seismic hazards***. Although the likelihood of simultaneous flood and seismic hazards is small, each hazard should be identified carefully and factored into siting, design, and construction practices.

Figure 2-11.
Galveston Island beach house with wind damage to roof in high pressure zones at roof edge and roof corners, Hurricane Ike, 2008

2.3.2 Siting

There is inherent risk in building near a coast, but this risk can be reduced through proper siting practices. The effects of coastal storms and hurricanes on buildings provide regular lessons on the effects of siting in coastal environments.

> **CROSS REFERENCE**
>
> Chapter 4 discusses siting considerations, siting practices to avoid, and recommended alternatives.

Building close to the shoreline is a common, and often poor, siting practice. It generally renders a building more vulnerable to wave, flood, and erosion effects and reduces any margin of safety against multiple storms or erosion events. If flood hazards increase over time, the building may require removal, protection, or demolition. In coastal areas subject to long-term or episodic erosion, poor siting often leads to otherwise well-built elevated buildings standing on the active beach. While considered a structural success, such buildings are generally uninhabitable because of the loss of utilities and access. The presence of homes on active beaches can also lead to conflicts over beach use and increase pressure to armor or re-nourish beaches (both controversial and expensive measures). ***Buildings sited on naturally occurring rocky shorelines*** are better protected from erosion and direct wave impacts, but may still be subject to wave overtopping.

Buildings subject to storm-induced erosion, including those in low-lying areas and buildings ***sited on the tops of erodible dunes and bluffs*** are vulnerable to damage caused by the undermining of foundations and the loss of supporting soil around vertical foundation members. Building on dunes and bluffs is discouraged. If buildings are constructed on dunes or bluffs they must be sited far from erodible slopes and must have a deep, well-designed, and well-constructed pile or column foundation.

CROSS REFERENCE

Figures 3-37 and 3-46 show the consequences of siting buildings on the tops of erodible bluffs.

The additional hazards associated with building near naturally occurring geographic features should be considered. Siting along shorelines protected against wave attack by barrier islands or other land masses does not guarantee protection from flooding. In fact, ***storm surge elevations along low-lying shorelines in embayments*** are often higher than storm surge elevations on open coast shorelines. ***Buildings sited near unstabilized tidal inlets*** or in areas subject to large-scale shoreline fluctuations may be vulnerable to even minor storms or erosion events.

Building close to other structures may increase the potential for damage from flood, wind, debris, and erosion hazards. Siting homes or other small buildings adjacent to large, engineered high-rise structures is a particular concern. The larger structures can redirect and concentrate flood, wave, and wind forces, and have been observed to increase flood and wind forces, as well as scour and erosion, to adjacent structures. ***Siting near erosion control or flood protection structures*** has contributed to building damage or destruction because these structures may not afford the required protection during a design event. Seawalls, revetments, berms, and other structures may themselves be vulnerable as a result of erosion and scour or other prior storm impacts. Siting too close to protective structures may preclude or make difficult any maintenance of the protective structure. ***Buildings sited on the downdrift shoreline*** of a groin or stabilized tidal inlet (an inlet whose location has been fixed by jetties) may be subject to increased erosion. Figure 2-12 shows how increased erosion rates on the downdrift side of groins can threaten structures.

Building in a levee-impacted area has special risks that should be understood. Levees are common flood protection structures in some coastal areas. The purpose of a levee is to reduce risk from temporary flooding to the people and property behind it (known as levee-impacted areas). Levees are designed to provide a specific level of risk reduction (e.g., protection from the 1-percent-annual-chance flood). It must be remembered that levees can be overtopped or breached during floods that are larger than they were designed to withstand. Levees can also fail during floods that are less than the design level due to inadequacies in design, construction, operation, or maintenance.

TERMINOLOGY: LEVEE

A levee is a man-made structure, usually an earthen embankment, built parallel to a waterway to contain, control, or divert the flow of water. A levee system may also include concrete or steel floodwalls, fixed or operable floodgates and other closure structures, pump stations for rainwater drainage, and/or other elements, all of which must perform as designed to prevent failure.

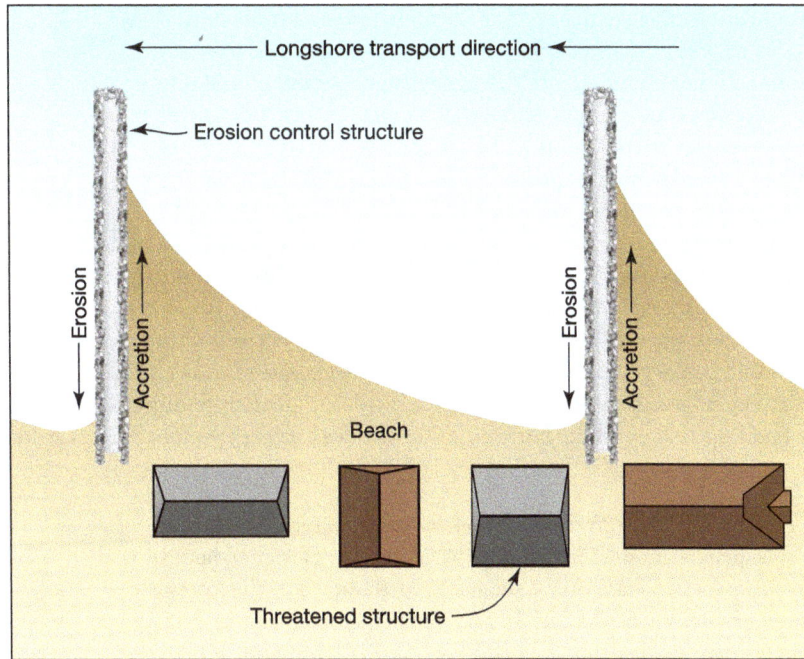

Figure 2-12.
Structures built close
to the downdrift side of
groins and jetties can
experience increased
erosion rates

SOURCE: ADAPTED FROM
MAINE GEOLOGICAL SURVEY
2005

When levees fail, it is often catastrophic. In 2005, Hurricane Katrina's storm surge caused failure of the certified levee system protecting New Orleans, LA, and flooded almost 80 percent of the city, making Hurricane Katrina the most destructive natural disaster in the history of the United States. The flooding was caused by a combination of breaching and overtopping. Flood levels were higher than the BFE for most of the affected area, rising well above the first floor, even for buildings elevated above the BFE.

An additional hazard related to levee overtopping or breaching is that resultant flooding may have a much longer duration, perhaps as long as a few weeks, compared to that of coastal floods, which typically last a day or less. Long-duration floods can increase damage to buildings through mold growth, corrosion, and other deterioration of building materials.

CROSS REFERENCE

Section 3.6.9 discusses NFIP treatment of levees.

No levee is flood-proof, and regular inspection, maintenance, and periodic upgrades of levees are necessary to maintain the desired level of protection. Homeowners sited behind levees should take precautions, such as elevating and floodproofing their homes, and be prepared to evacuate in an emergency. For more information, refer to *So, You Live Behind a Levee!* (ASCE 2010b).

2.3.3 Design

Building design is one of the most important factors of a successful coastal building. Observations of building damage resulting from past storm events have not only provided insight into the design of coastal buildings, but have led to positive changes in building design codes and standards. Newer buildings built to these codes tend to perform better. However, certain design flaws still exist and are observed year after year.

Foundation design is an important factor in the success of a coastal building. Use of *shallow spread footing and slab foundations* in areas subject to wave impact and/or erosion can result in building collapse, even during minor flood or erosion events. Because of the potential for undermining by erosion and scour, this type of foundation may not be appropriate for coastal bluff areas outside the mapped floodplain and some Coastal A Zones. Figure 2-13 shows an extreme case of localized scour undermining a slab-on-grade house after Hurricane Fran. The lot was mapped as Zone A and located several hundred feet from the shoreline. This case illustrates the need for open foundations in Coastal A Zones. Use of *continuous perimeter wall foundations*, such as crawlspace foundations (especially unreinforced masonry) in areas subject to wave impact and/or erosion may result in building damage, collapse, or total loss. For *open foundations*, inadequate depth of foundation members is a common cause of failure in pile-elevated one- to four-family residential buildings. Figure 2-14 shows a building that survived Hurricane Katrina with a deeply embedded pile foundation that is sufficiently elevated.

In addition, *insufficient elevation* of a building exposes the superstructure to damaging wave forces. Designs should incorporate freeboard above the required elevation of the lowest floor or bottom of the lowest horizontal member. Figure 2-15 shows two neighboring homes. The pre-FIRM house on the left experienced significant structural damage due to surge and waves. The newer, post-FIRM house on the right sustained minor damage because it was elevated above grade, and grade had been raised a few feet by fill.

In addition to foundation design, there are other commonly observed points of failure in the design of coastal buildings. *Failure to provide a continuous load path* from the roof to the foundation using adequate connections may lead to structural

CROSS REFERENCE

Chapter 10 provides a detailed discussion of foundation design.

TERMINOLOGY: LOWEST FLOOR

Under the NFIP, the "lowest floor" of a building includes the floor of a basement. The NFIP regulations define a basement as "... any area of a building having its floor subgrade (below ground level) on all sides." For insurance rating purposes, this definition applies even when the subgrade floor is not enclosed by full-height walls.

Figure 2-13.
Extreme case of localized scour undermining a Zone A continuous perimeter wall foundation located several hundred feet from the shoreline, Hurricane Fran (Topsail Island, NC, 1996)

Figure 2-14.
Successful example of well-elevated and embedded pile foundation tested by Hurricane Katrina. Note adjacent building failures (Dauphin Island, AL, 2005)

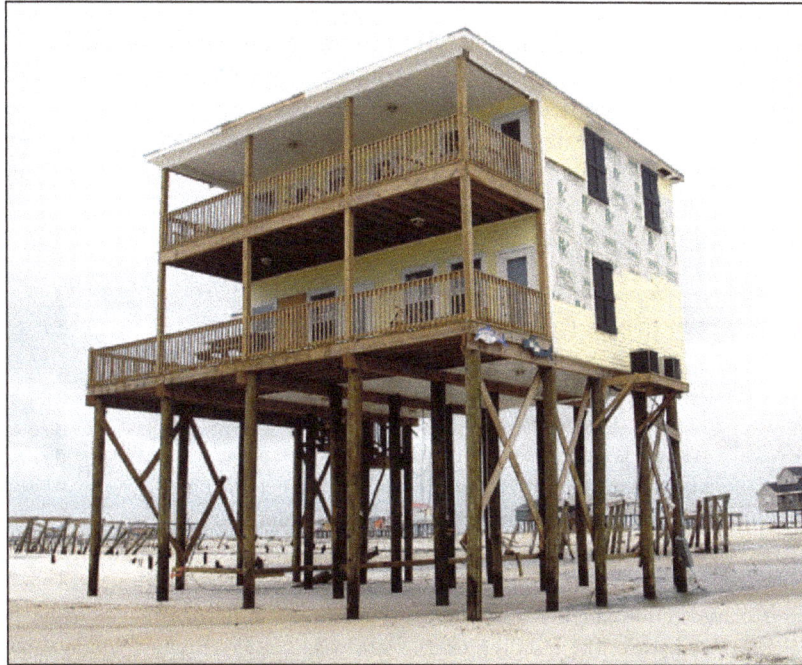

Figure 2-15.
The pre-FIRM house (left) experienced damage due to surge and waves while the newer, elevated, post-FIRM house (right) experienced minimal damage, Hurricane Ivan (Santa Marina, Pensacola, FL, 2005)

failure. ***Failure to use corrosion-resistant structural connectors*** can compromise structural integrity and may lead to building failures under less than design conditions. Examples of corrosion-resistant connectors include wooden connectors, heavy gauge galvanized connectors, and stainless steel connectors. Salt spray and breaking waves accelerate

CROSS REFERENCE

Chapter 9 includes discussion on designing a continuous load path.

Section 9.2.3 discusses connectors.

corrosion of metal building components. Nails, screws, sheet-metal connector straps, and truss plates made of ferrous metals are the most likely to corrode. ***Decks and roofs*** supported by inadequately embedded vertical members, especially those that are multiple stories, can lead to major structural damage even during minor flood and erosion events. Failure to adequately connect ***porch roofs*** and to limit the size of ***roof overhangs*** can lead to extensive damage to the building envelope during minor wind events. Roof overhangs should be designed to remain intact without vertical supports. Alternatively, supports should be designed to the same standards as the main foundation. Decks must be designed to withstand all design loads or should be designed so that they do not damage the main building when they fail.

Building envelopes are susceptible to wind damage, wind debris, and water penetration. Protection of the entire building envelope is necessary in high-wind areas. It is recommended that glazing in hurricane-prone areas be protected; however, in wind-borne debris regions as defined by the governing building code and ASCE-7, glazing is required to be protected by temporary or permanent storm shutters or impact-resistant glass. In addition to preventing pressurization, opening protection will reduce damage caused by wind, wind-borne debris, and rainfall penetration.

CROSS REFERENCE

Chapter 11 provides a detailed discussion of building envelope design, including exterior walls, windows, doors, and roofs.

However, proper specification of windows, doors, and their attachment to the structural frame is essential for full protection. Figure 2-16 shows two similar buildings in the same neighborhood that survived Hurricane Charley. The building on the left lost its roof structure due to internal pressurization resulting from unprotected windows and doors. The building on the right was protected with shutters and the roof sustained relatively minor damage.

Many ***commonly used residential roofing designs***, techniques, systems, and materials are susceptible to damage from wind and wind-borne debris. Designers should carefully consider the selection and attachment of roof sheathing and roof coverings in coastal areas. ***Low-slope roofs*** may experience higher wind loads and must effectively drain the heavy rains accompanying coastal storms. As with all houses, the designer should

Figure 2-16.
The unprotected building sustained roof damage due to pressurization (left) while the other sustained only minor damage because it was protected by shutters (right), Hurricane Charley (Captiva Island, FL, 2004)

ensure that all loads, drainage, and potential water infiltration problems are addressed. Roof designs that incorporate **gable ends** (especially those that are unbraced) and **wide overhangs** are susceptible to failure (Figure 2-17) unless adequately designed and constructed for the expected loads. Alternative designs that are more resistant to wind effects should be used in coastal areas.

The design and placement of **swimming pools** can affect the performance of adjacent buildings. In-ground and above-ground (but below the DFE) pools should not be structurally attached to buildings. An attached pool can transfer flood loads to the building. Building foundation designs should also account for the effects of non-attached but adjacent pools: increased flow velocities, wave runup, wave reflection, and scour that can result from the redirection of flow by the pool. In addition, swimming pools should not be installed in enclosures below elevated buildings.

Figure 2-17.
Wind damage to roof structure and gable end wall, Hurricane Katrina (Pass Christian, MS, 2005)

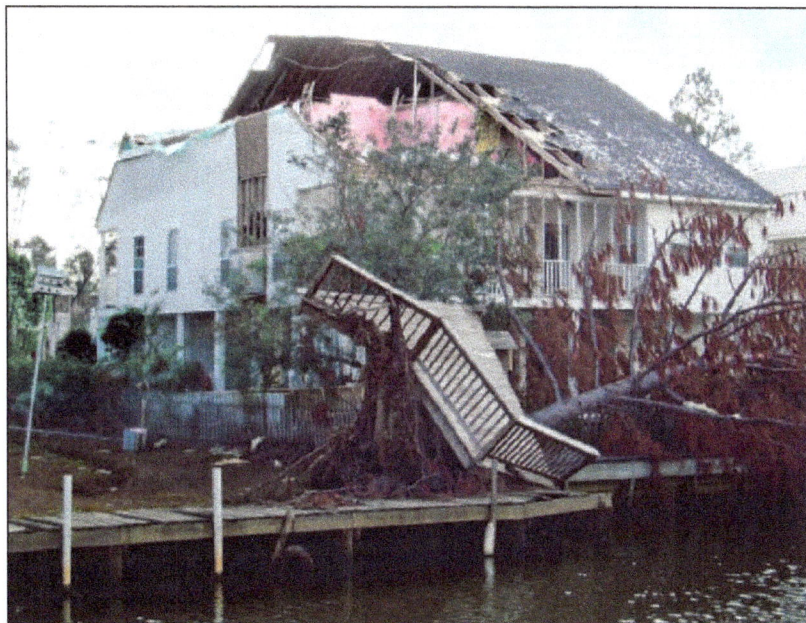

2.3.4 Construction

Post-disaster observations often indicate that damage could have been reduced if buildings had been constructed according to approved designs and using best practices. Careful preparation of design documents and attention to construction details can reduce damage to coastal homes. FEMA P-499, *Home Builder's Guide to Coastal Construction* (FEMA 2010) and the NFIP Technical Bulletin Series Numbers 1 through 11 (FEMA 1993-

CROSS REFERENCE

Chapter 13 provides details on construction of coastal buildings.

2011, available at http://www.fema.gov/plan/prevent/floodplain/techbul.shtm) provide detailed technical guidance and recommendations concerning the construction of coastal residential buildings.

Failure to achieve the pile or foundation embedment specified by building plans or local and State requirements will render an otherwise properly constructed building vulnerable to flood, erosion, and scour damage. *Improperly constructed breakaway walls* (e.g., improperly fastened wall panels or panels constructed immediately seaward of foundation cross-bracing) can cause preventable damage to the main structure during a flood event.

Poorly made structural connections, particularly in wood frame and masonry structures, (e.g., pile/pier/ column-to-beam, joist-to-beam) have caused the failure of residential structures throughout the coastal areas of the United States. Proper embedment and lap splicing of reinforcing in concrete piers and footings is critical. Figure 2-18 shows an example of a masonry column connection that failed during Hurricane Katrina. Post-event investigations have revealed many instances of *inadequate connections* (e.g., improper or inadequately sized fasteners) that either failed during the event or could have failed if the design loads had been realized at the connection. Connections must be made with the appropriate fastener for the design structural capacity. Nail guns, frequently used to speed construction, can easily *over drive nails*, or drive them at an angle, leading to connections with reduced capacity. In addition, the nail gun operator may not be able to determine whether the nail has penetrated an unexposed wood member as intended, such as for a rafter or truss below the roof sheathing. Staples are not appropriate for connecting wood members in coastal areas.

Bracing and fastening roofs and walls can help prevent building envelope failures in high-wind events. While bracing and fastening is adequately addressed in most current codes, older buildings built to older codes may be constructed with *inadequate bracing and fastening*. Lack of, or *inadequate, connections between shingles and roof sheathing* and between sheathing and roof framing (e.g., nails that fail to penetrate roof truss members or rafters) can cause roof failures and subsequent building failures.

Figure 2-18.
Failed masonry column connection, Hurricane Katrina (Jackson County, MS, 2005)

2.3.5   Enclosures

Enclosures present a unique situation to coastal construction. NFIP regulations state that the area below an elevated building can be used only for parking, building access, and storage. These areas must not be finished or used for recreational or habitable purposes. No mechanical, electrical, or plumbing equipment is to be installed below the BFE. However, post-construction conversion of enclosures to habitable space remains a common violation of floodplain management requirements and is difficult for communities and States to control.

Designers and owners should realize that: (1) enclosures and items in them are likely to be damaged or destroyed even during minor flood events; (2) enclosures, and most items in them, are not covered by flood insurance and, if damaged, the owner may incur significant costs to repair or replace them; and (3) even if enclosures are properly constructed with breakaway walls, the presence of enclosures increases flood insurance premiums for the entire building (the premium rate increases with the size of the enclosed area). Therefore, enclosed areas below elevated buildings, even if compliant with NFIP design and construction requirements, can have significant future cost implications for homeowners.

Enclosures can have two types of walls:

- ***Enclosures with breakaway walls*** are designed to collapse under flood loads and act independently from the elevated building, leaving the foundation intact (Figure 2-19). All enclosures below elevated buildings in Zone V must have breakaway walls. Enclosures in Zone A and Coastal A Zones may have breakaway walls, but the walls must have flood openings to comply with Zone A requirements.

- ***Enclosures and closed foundations that do not have breakaway walls*** can be constructed below elevated buildings in Zone A but are not recommended in Coastal A Zones. The walls of enclosures and foundation walls below elevated buildings in Zone A must have flood openings to allow the free entry and exit of floodwaters (Figure 2-20).

Taller breakaway walls appear to produce larger pieces of flood-borne debris. Post-disaster investigations have observed some breakaway walls in excess of 11 feet high (FEMA 2009b). These investigations have also observed that louvered panels (Figure 2-21) remained intact longer than solid breakaway walls under the same flood conditions. As a result, houses with louvered panels had less flood-related damage (and repair cost) and generated less flood-borne debris. The use of louver panels can also result in lower flood insurance

TERMINOLOGY: ENCLOSURE

An enclosure is formed when any space below the lowest floor is enclosed on all sides by walls or partitions.

CROSS REFERENCE

Section 9.3 discusses the proper design of breakaway walls.

NOTE

A change beginning with the May 2009 FEMA *Flood Insurance Manual* rates Zone V enclosures as "free of obstructions" if they are constructed with louvers or lattice on all walls except one (for garage door or solid breakaway wall). Previous rating practice called this "with obstruction."

Figure 2-19.
Breakaway walls below
the first floor of this
house broke as intended
under the flood forces
of Hurricane Ike (Bolivar
Peninsula, TX, 2008)

Figure 2-20.
Flood opening in an
enclosure with breakaway
walls, Hurricane Ike
(Galveston Bay shoreline,
San Leon, TX)

premiums. Flood insurance premiums for a building located in Zone V are much less when a below-BFE enclosure is formed by louvers than by breakaway walls. A building with an enclosure formed by louvers is classified the same as if it had insect screening or open lattice (Figure 2-22), i.e., as "free of obstructions," while a solid breakaway wall enclosure results in a "with obstruction" rating for the building.

Figure 2-21.
Louvers installed beneath an elevated house are a good alternative to breakaway walls
SOURCE: FEMA P-499 2010

Figure 2-22.
An enclosure formed by open lattice (Isle of Palms, SC)

Two other enclosure scenarios have design and flood insurance implications. Designers should be cautious when an owner asks for either type of enclosure, and should consult with the community and a knowledgeable flood insurance agent:

- **Enclosures that do not extend all the way to the ground (sometimes called "above-grade," "hanging," or "elevated" enclosures).** These enclosures have a floor system that is not in contact with the ground, but that may be connected to the building foundation or supported on the primary pile system or short posts (Figure 2-23). Having the floor of the enclosure above grade means frequent flooding passes underneath, which may reduce the frequency and severity of damage. These enclosures were not contemplated when flood insurance premium rate tables were prepared, and thus can result in significantly higher flood insurance premiums. As of early 2011, the NFIP was working to address this type of construction, but until such time as it is resolved, owners will pay a substantial premium penalty for this type of enclosure.

- **Two-story enclosures.** In flood hazard areas with very high BFEs, some owners have constructed two-story, solid walls to enclose areas below elevated buildings, typically with a floor system approximately midway between the ground and the elevated building (Figure 2-24). These enclosures present unique problems. In Zone A, the walls at both levels of the enclosure must have flood openings; there must be some means to relieve water pressure against the floor system between the upper and lower enclosures; and special ingress and egress code requirements may apply. These enclosures may also result in substantially higher flood insurance premiums.

Figure 2-23.
Above-grade enclosure
(Perry, FL)

Figure 2-24.
Two-story enclosure
SOURCE: FEMA P-499 2010

2.3.6 Maintenance

Repairing and replacing structural elements, connectors, and building envelope components that have deteriorated because of decay or corrosion helps to maintain a building's resistance to natural hazards. Maintenance of building components in coastal areas should be an ongoing process. The ultimate costs of deferred maintenance in coastal areas can be high when natural disasters strike. *Failure to inspect and repair damage* caused by wind, flood, erosion, or other hazard makes a building even more vulnerable during the next event. *Failure to maintain erosion control or coastal flood protection structures* leads to increased vulnerability of those structures and the buildings behind them.

CROSS REFERENCE

Chapter 14 provides details on the maintenance of coastal buildings.

2.4 References

ASCE (American Society of Civil Engineers). 1998. *Minimum Design Loads for Buildings and Other Structures.* ASCE Standard ASCE 7-98.

ASCE. 2010a. *Minimum Design Loads for Buildings and Other Structures.* ASCE Standard ASCE 7-10.

ASCE. 2010b. *So, You Live Behind a Levee!*

California Coastal Commission. 1997. *Questions and Answers on El Nino.* http://www.coastal.ca.gov/elnino/enqa.html. Accessed 03/02/11.

Eefting, T. 1927. "Structural Lessons of the South Florida Hurricane." Florida Engineer and Contractor. September. pp. 162–170.

FEMA (Federal Emergency Management Agency). 1980. *Elevating to the Wave Crest Level — A Benefit-Cost Analysis.* FIA-6.

FEMA. 1992. *Building Performance Assessment Team, Field Trip and Assessment within the States of Maryland and Delaware in Response to a Nor'easter Coastal Storm on January 4, 1992.* Final Report. March 4.

FEMA. 1993. *Building Performance: Hurricane Andrew in Florida, Observations, Recommendations and Technical Guidance.* FIA-22.

FEMA.1996. *Hurricane Opal in Florida, A Building Performance Assessment.* FEMA-281.

FEMA. 1997. *Building Performance Assessment: Hurricane Fran in North Carolina, Observations, Recommendations and Technical Guidance.* FEMA-290.

FEMA. 1999a. *Hurricane Georges in the Gulf Coast – Observations, Recommendations, and Technical Guidance.* FEMA 338.

FEMA. 1999b. *Hurricane Georges in Puerto Rico – Observations, Recommendations, and Technical Guidance.* FEMA 339.

FEMA. 2005a. *Hurricane Charley in Florida – Observations, Recommendations, and Technical Guidance.* FEMA 488.

FEMA. 2005b. *Hurricane Ivan in Alabama and Florida – Observations, Recommendations, and Technical Guidance.* FEMA 489.

FEMA. 2006. *Hurricane Katrina in the Gulf Coast – Building Performance Observations, Recommendations, and Technical Guidance.* FEMA 549.

FEMA. 2009a. *Protecting Manufactured Homes from Floods and Other Hazards – A Multi-Hazard Foundation and Installation Guide.* FEMA P-85, Second Edition.

FEMA. 2009b. *Hurricane Ike in Texas and Louisiana – Building Performance Observations, Recommendations, and Technical Guidance.* FEMA P-757.

FEMA. 2010. *Home Builder's Guide to Coastal Construction Technical Fact Sheet Series.* FEMA P-499.

FEMA. 2011. *National Flood Insurance Program, Flood Insurance Manual.*

Hayes, M. O. 1967. "Hurricanes as Geological Agents: Case Studies of Hurricanes Carla, 1961, and Cindy, 1963." *Bureau of Economic Geology Report of Investigation No. 61.* Austin, TX: University of Texas.

Jibson, R. 2005. *Landslide Hazards at La Conchita, California.* Prepared for the United States Geological Survey. Open-File Report 2005-1067.

Maine Geological Survey. 2005. *Coastal Marine Geology: Frequently Asked Questions.* http://www.maine.gov/doc/nrimc/mgs/explore/marine/faq/groins.htm. Accessed 11/24/2010.

McGrath, J., ed. 1985. "California's Battered Coast." Proceedings from a February 6–8, 1985, Conference on Coastal Erosion. California Coastal Commission.

Minsinger, W.E. 1988. *The 1938 Hurricane, An Historical and Pictorial Summary.* Blue Hill Meteorological Observatory, East Milton, MA. Greenhills Books, Randolph Center, VT.

National Academy of Sciences, National Research Council, Commission on Engineering and Technical Systems. 1984. *Hurricane Alicia, Galveston and Houston, Texas, August 17–18, 1983.*

NOAA (National Oceanic and Atmospheric Administration). 2004. *Effects of Hurricane Isabel on Water Levels: Data Report.* NOS CO-OPS 040.

National Roofing Contractors Association. 1996. *Hurricane Marilyn, Photo Report of Roof Performance.* Prepared for the Federal Emergency Management Agency. March.

RMS (Risk Management Solutions). 2001. *Tropical Storm Allison, June 2001: RMS Event Report.*

Thom, H. C. S.; R. D. Marshall. 1971. "Wind and Surge Damage due to Hurricane Camille." *ASCE Journal of Waterways, Harbors and Coastal Engineering Division.* May. pp. 355–363.

TTU (Texas Tech University). 1990. *Performance of roofing systems in Hurricane Hugo.* August.

URS Group, Inc. (URS). 1989. *Flood Damage Assessment Report: Sandbridge Beach, Virginia, and Nags Head, North Carolina, April 13, 1988, Northeaster.* Prepared for the Federal Emergency Management Agency. March.

URS. 1990. *Flood Damage Assessment Report: Nags Head, North Carolina, Kill Devil Hills, North Carolina, and Sandbridge Beach, Virginia, March 6–10, 1989, Northeaster.* Prepared for the Federal Emergency Management Agency. April.

URS. 1991a. *Flood Damage Assessment Report: Surfside Beach to Folly Island, South Carolina, Hurricane Hugo, September 21–22, 1989.* Volume I. Prepared for the Federal Emergency Management Agency. August.

URS. 1991b. *Follow-Up Investigation Report: Repair Efforts 9 Months After Hurricane Hugo, Surfside Beach to Folly Island, South Carolina.* Volume I. Prepared for the Federal Emergency Management Agency. August.

URS. 1991c. *Flood Damage Assessment Report: Buzzard's Bay Area, Massachusetts, Hurricane Bob, August 19, 1991.* October.

USACE (U.S. Army Corps of Engineers). 1997. "Annual Summary." *Great Lakes Update.* Vol. No. 126. Detroit, MI: Detroit District. January 3.

USACE. 1998. "Annual Summary." *Great Lakes Update.* Detroit, MI: Detroit District.

Waldon, D. 1990. "Raising Galveston." *American Heritage of Invention and Technology.* Vol. 5, No. 3, pp. 8–18.

Wilson, B. W. and A. Torum. 1968. *The Tsunami of the Alaskan Earthquake, 1964: Engineering Evaluation.* Technical Memorandum No. 25. U.S. Army Corps of Engineers, Coastal Engineering Research Center.

York, Michael. "Deadly Hugo Slams Puerto Rico, Virgin Islands." *Washington Post.* September 19, 1989.

Identifying Hazards

Buildings constructed in coastal areas are subject to natural hazards. The most significant natural hazards that affect the coastlines of the United States and territories can be divided into four general categories:

- Coastal flooding (including waves)
- Erosion
- High winds
- Earthquakes

This chapter addresses each of these categories, as well as other hazards and environmental effects, but focuses on flooding and erosion (Sections 3.4 and 3.5). These two hazards are among the least understood and the least discussed in design and construction documents. Designers have numerous resources available that discuss wind and seismic hazards in detail, so they will be dealt with in less detail here.

In order to construct buildings to resist these natural hazards and reduce existing buildings' vulnerability to such hazards, proper planning, siting, design, and construction are critical and require an understanding of the coastal environment, including coastal geology, coastal processes, regional variations in coastline characteristics, and coastal sediment budgets. Proper siting and design also require accurately assessing the

CROSS REFERENCE

For resources that augment the guidance and other information in this Manual, see the Residential Coastal Construction Web site (http://www.fema.gov/rebuild/mat/fema55.shtm).

WARNING

Natural hazards can act individually, but often act in combination (e.g., high winds and coastal flooding, coastal flooding and erosion, etc.). Long-term changes in underlying conditions—such as sea level rise—can magnify the adverse effects of some of these hazards. For more information on load combinations, see Chapter 8.

vulnerability of any proposed structure, including the nature and extent of its exposure to coastal hazards. Failure to properly identify and design to resist coastal hazards expected over the life of a building can lead to severe consequences, most often building damage or destruction.

This chapter provides an overview of coastline characteristics (Section 3.1); tropical cyclones and coastal storms (Section 3.2); coastal hazards (Section 3.3); coastal flood effects, including erosion (Sections 3.4 and 3.5); and flood hazard zones and assessments, including hazard mapping procedures used by the NFIP (Sections 3.6 and 3.7). Although general guidance on identifying hazards that may affect a coastal building site is provided, this chapter does not provide specific hazard information for a particular site. Designers should consult the sources of information listed in Chapter 4 of this Manual and in the resource titled "Information about Storms, Big Waves, and Water Levels" on the FEMA Residential Coastal Construction Web page. Siting considerations are discussed in more detail in Chapter 4.

3.1   Coastline Characteristics

This section contains general information on the coastal environment and the characteristics of the United States coastline.

3.1.1   Coastal Environment

Coastal geology and geomorphology refer to the origin, structure, and characteristics of the rocks and sediments that make up the coastal region. The coastal region is considered the area from the uplands to the nearshore as shown in Figure 3-1. Coastal sediments can vary from small particles of silt or sand (a

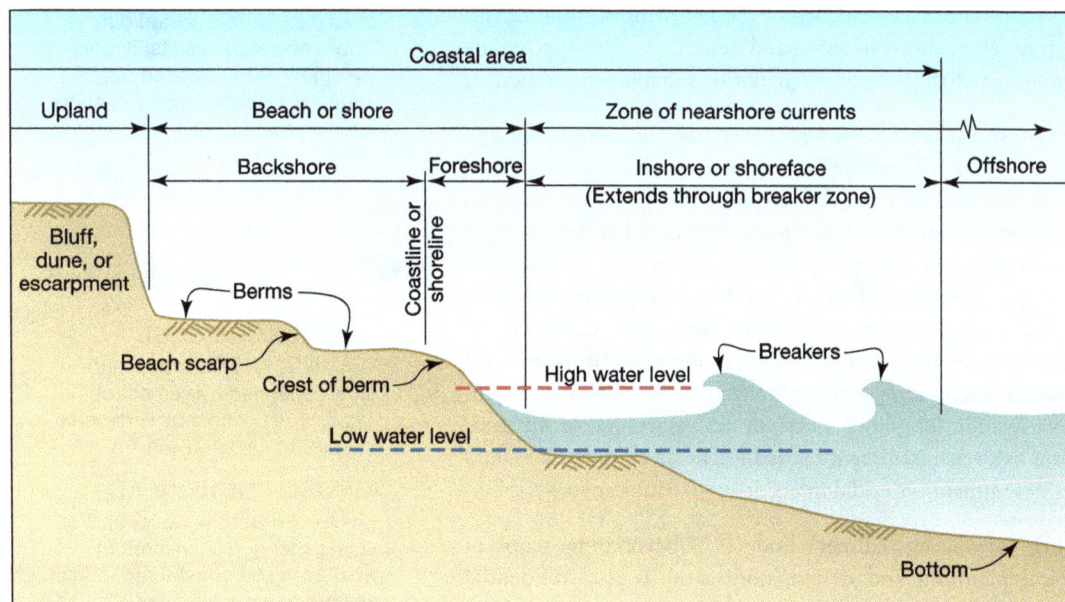

Figure 3-1. Coastal region terminology
SOURCE: ADAPTED FROM USACE 2008

few thousandths or hundredths of an inch across), to larger particles of gravel and cobble (up to several inches across), to formations of consolidated sediments and rock. The sediments can be easily erodible and transportable by water and wind, as in the case of silts and sands, or can be highly resistant to erosion. The sediments and rock units that compose a coastline are the product of physical and chemical processes that take place over thousands of years.

Coastal processes refer to physical processes that act upon and shape the coastline. These processes, which influence the configuration, orientation, and movement of the coast, include the following:

- Tides and fluctuating water levels
- Waves
- Currents (usually generated by tides or waves)
- Winds

Coastal processes interact with the local coastal geology to form and modify the physical features that are referred to frequently in this Manual: beaches, dunes, bluffs, and upland areas. Water levels, waves, currents, and winds vary with time at a given location (according to short-term, seasonal, or longer-term patterns) and vary geographically at any point in time. A good analogy is weather; weather conditions at a given location undergo significant variability over time, but tend to follow seasonal and other patterns. Further, weather conditions can differ substantially from one location to another at the same point in time.

Regional variations in coastlines are the product of variations in coastal processes and coastal geology. These variations can be quite substantial, as described in the following sections of this chapter. Thus, shoreline siting and design practices appropriate to one area of the coastline may not be suitable for another.

The *coastal sediment budget* is based on the identification of sediment sources and sinks, and refers to the quantification of the amounts and rates of sediment transport, erosion, and deposition within a defined region. Sediment budgets are used by coastal engineers and geologists to analyze and explain shoreline changes and to project future shoreline behavior. Typical sediment sources include longshore transport of sediment into an area, beach nourishment, and dune or bluff erosion (which supply sediment to the beach). Typical sediment sinks include longshore sediment transport out of an area, storm overwash (sediment carried inland from the beach), and loss of sediment into tidal inlets or submarine canyons.

While calculating sediment budgets is beyond the scope of typical planning and design studies for coastal residential structures, sediment budgets may have been calculated by others for the shoreline segment containing a proposed building

NOTE

Although calculating coastal sediment budgets can be complicated, the premise behind it is simple: if more sediment is transported by coastal processes or human actions into a given area than is transported out, shore accretion results; if more sediment is transported out of an area than is transported in, shore erosion results.

TERMINOLOGY

LONGSHORE SAND TRANSPORT is wave- and/or tide-generated movement of shallow-water coastal sediments parallel to the shoreline.

CROSS-SHORE SAND TRANSPORT is wave- and/or tide-generated movement of shallow-water coastal sediments toward or away from the shoreline.

site. Designers should contact State coastal management agencies and universities to determine if sediment budget and shoreline change information for their site is available, since this information will be useful in site selection, planning, and design.

The concept of sediment budgets does not apply to all coastlines, particularly rocky coastlines that are resistant to erosion and whose existence does not depend on littoral sediments transported by coastal processes. Rocky coastlines typical of many Pacific, Great Lakes, New England, and Caribbean areas are better represented by Figure 3-2. The figure illustrates the slow process by which rocky coasts erode in response to elevated water levels, waves, and storms.

3.1.2 United States Coastline

The estimated total shoreline length of the continental United States, Alaska, and Hawaii is 84,240 miles, including 34,520 miles of exposed shoreline and 49,720 miles of sheltered shoreline (USACE 1971). The shoreline length of the continental United States alone is estimated as 36,010 miles (13,370 miles exposed, 22,640 miles sheltered).

Several sources (National Research Council 1990, Shepard and Wanless 1971, USACE 1971) were used to characterize and divide the coastline of the United States into six major segments and several smaller subsegments (see Figure 3-3). Each of the subsegments includes coastlines of similar origin, characteristics, and hazards.

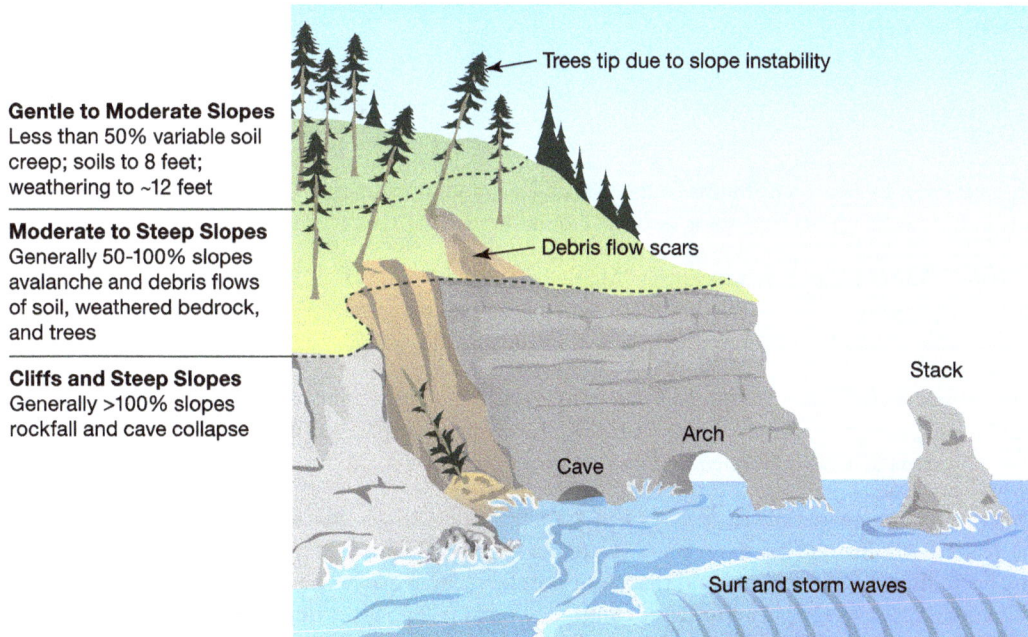

Figure 3-2.
Generalized depiction of erosion process along a rocky coastline
SOURCE: ADAPTED FROM HORNING GEOSCIENCES 1998

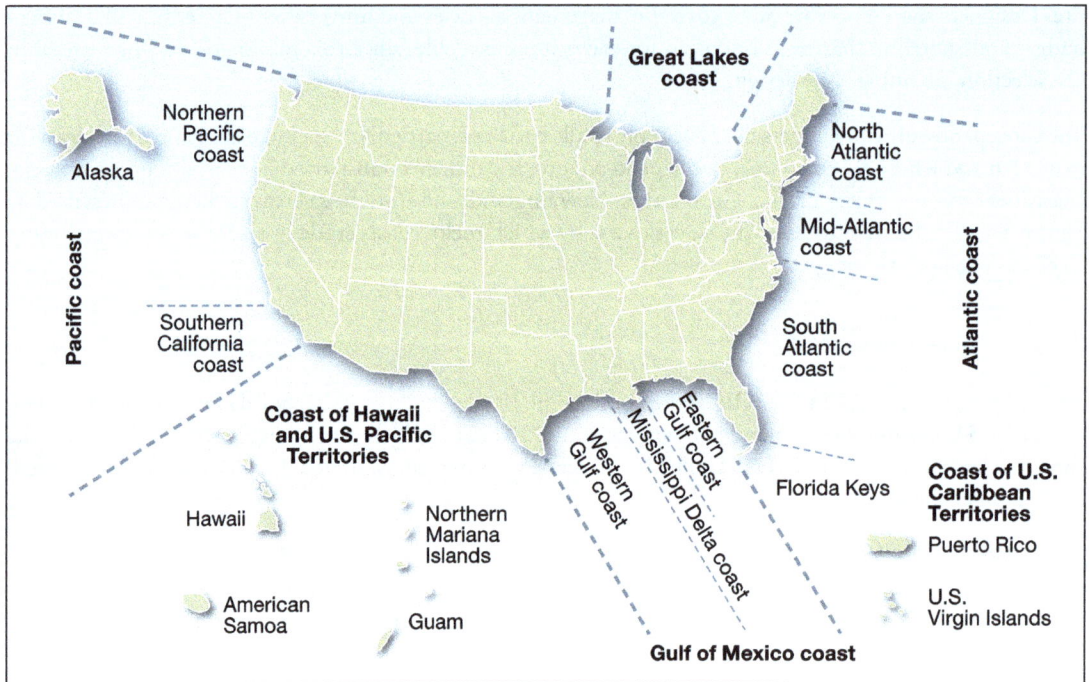

Figure 3-3.
United States coastline

Atlantic Coast

The Atlantic coast extends from Maine to the Florida Keys and includes the North Atlantic coast, the Mid-Atlantic coast, the South Atlantic coast, and the Florida Keys.

The ***North Atlantic coast***, extending from Maine to Long Island, NY, is glacial in origin. It is highly irregular, with erosion-resistant rocky headlands and pocket beaches in northern New England, and erodible bluffs and sandy barrier islands in southern New England and along Long Island, NY.

The ***Mid-Atlantic coast*** extends from New Jersey to Virginia, and includes two of the largest estuaries in the United States; Delaware Bay and Chesapeake Bay. The open coast shoreline is generally composed of long barrier islands separated by tidal inlets and bay entrances.

The ***South Atlantic coast*** extends from North Carolina to South Florida and consists of three regions: (1) the North Carolina and northern South Carolina shoreline, composed of long barrier and mainland beaches (including the Outer Banks and the South Carolina Grand Strand region); (2) the region extending from Charleston, SC, to the St. Johns River entrance at Jacksonville, FL (a tide-dominated coast composed of numerous short barrier islands, separated by large tidal inlets and backed by wide expanses of tidal marsh); and (3) the east coast of Florida (composed of barrier and mainland beaches backed by narrow bays and rivers).

The **Florida Keys** are a series of low-relief islands formed by limestone and reef rock, with narrow, intermittent carbonate beaches.

The entire Atlantic coast is subject to waves and high storm surges from hurricanes and/or nor'easters. Wave runup on steeply sloping beaches and shorelines in New England is also a common source of coastal flooding.

Gulf of Mexico Coast

The Gulf of Mexico coast extends from the Florida Keys to Texas. It can be divided into three regions: (1) the **eastern Gulf Coast** from southwest Florida to Mississippi, which is composed of low-lying sandy barrier islands south of Tarpon Springs, FL, and west of St. Marks, FL, with a marsh-dominated coast in between in the Big Bend area of Florida; (2) the **Mississippi Delta Coast** of southeast Louisiana, characterized by wide, marshy areas and a low-lying coastal plain; and (3) the **western Gulf Coast**, including the cheniers of southwest Louisiana, and the long, sandy barrier islands of Texas.

The entire Gulf of Mexico coast is vulnerable to high storm surges and waves from hurricanes. Some areas (e.g., the Big Bend area of Florida) are especially vulnerable because of the presence of a wide, shallow continental shelf and low-lying upland areas.

Coast of U.S. Caribbean Territories

The islands of Puerto Rico and the U.S. Virgin Islands are the products of ancient volcanic activity. The coastal lowlands of Puerto Rico, which occupy nearly one-third of the island's area, contain sediment eroded and transported from the steep, inland mountains by rivers and streams. Ocean currents and wave activity rework the sediments on pocket beaches around each island. Coastal flooding is usually due to hurricanes, although tsunami events are not unknown in the Caribbean.

Great Lakes Coast

The shorelines of the Great Lakes coast extend from Minnesota to New York. They are highly variable and include wetlands, low and high cohesive bluffs, low sandy banks, and lofty sand dunes perched on bluffs (200 feet or more above lake level). Storm surges along the Great Lakes are generally less than 2 feet except in small bays (2 to 4 feet) and on Lake Erie (up to 8 feet). Large waves can accompany storm surges. Periods of active erosion are triggered by heavy precipitation events, storm waves, rising lake levels, and changes in groundwater outflow along the coast.

Pacific Coast

The Pacific coast extends from California to Washington, and includes Alaska. It can be divided into three regions: (1) the **southern California coast**, which extends from San Diego County to Point Conception (Santa Barbara County), CA, and is characterized by long, sandy beaches and coastal bluffs; (2) the **northern Pacific coast,** which extends from Point Conception, CA, to Washington and is characterized by rocky cliffs, pocket beaches, and occasional long sandy barriers near river mouths; and (3) the **coast of Alaska**.

Open coast storm surges along the Pacific shoreline are generally small (less than 2 feet) because of the narrow continental shelf and deep water close to shore. However, storm wave conditions along the Pacific shoreline are severe, and the resulting wave runup can be very destructive. In some areas of the Pacific coast, tsunami flood elevations can be much higher than flood elevations associated with coastal storms.

The *coast of Alaska* can further be divided into two areas: (1) the southern coast, dominated by steep mountainous islands indented by deep fjords, and (2) the Bering Sea and Arctic coasts, backed by a coastal plain dotted with lakes and drained by numerous streams and rivers. The climate of Alaska and the action of ice along the shorelines set it apart from most other coastal areas of the United States.

Coast of Hawaii and U.S. Pacific Territories

The islands that make up Hawaii are submerged volcanoes; thus, the coast of Hawaii is formed by rocky cliffs and intermittent sandy beaches. Coastlines along the Pacific Territories are similar to those of Hawaii. Coastal flooding can be due to two sources: storm surges and waves from hurricanes or cyclones, and wave runup from tsunamis.

3.2   Coastal Storm Events

Tropical cyclones and coastal storms occur in varying strengths and intensities in all coastal regions of the United States and its territories. These storms are the primary source of the flood and wind damage that the recommendations of this Manual aim to reduce. Tropical cyclones and coastal storms include all storms associated with circulation around an area of atmospheric low pressure. When the storm origin is tropical and the circulation is closed, tropical storms, hurricanes, or typhoons result.

Tropical cyclones and coastal storms are capable of generating high winds, coastal flooding, high-velocity flows, damaging waves, significant erosion, and intense rainfall (see Figure 3-4). Like all flood events, they are also capable of generating and moving large quantities of water-borne sediments and floating debris. Consequently, the risk to improperly sited, designed, or constructed coastal buildings can be great.

Figure 3-4. Storm surge flooded this home in Ascension Parish, LA (Tropical Storm Allison, 2001)

One parameter not mentioned in the storm classifications described in the following sections—***storm coincidence with spring tides or higher than normal water levels***—also plays a major role in determining storm impacts and property damage. If a tropical cyclone or other coastal storm coincides with abnormally high water levels or with the highest monthly, seasonal, or annual tides, the flooding and erosion impacts of the storm are magnified by the higher water levels, to which the storm surge and wave effects are added.

> **CROSS REFERENCE**
>
> See Section 3.5.5 for a discussion of high water levels and sea level rise.

3.2.1.1   Tropical Cyclones

Tropical storms have 1-minute sustained winds averaging 39 to 74 miles per hour (mph). When sustained winds intensify to greater than 74 mph, the resulting storms are called ***hurricanes*** (in the North Atlantic basin or in the Central or South Pacific basins east of the International Date Line) or ***typhoons*** (in the western North Pacific basin).

> **NOTE**
>
> NOAA has detailed tropical storm and hurricane track information from 1848 to the present (http://csc.noaa.gov/hurricanes).

Hurricanes are divided into five classes according to the Saffir-Simpson Hurricane Wind Scale (SSHWS), which uses 1-minute sustained wind speed at a height of 33 feet over open water as the sole parameter to categorize storm damage potential (see Table 3-1). The SSHWS, which replaces the Saffir-Simpson Hurricane Scale, was introduced for the 2010 hurricane season to reduce confusion about the impacts associated with the hurricane categories and to provide a more scientifically defensible scale (there is not a strict correlation between wind speed and storm surge, as the original scale implied, as demonstrated by recent storms [e.g., Hurricanes Katrina and Ike] which produced devastating surge damage even though wind speeds at landfall were associated with lower hurricane categories). The storm surge ranges, flooding impact, and central pressure statements were removed from the original scale, and only peak wind speeds are included in the SSHWS (NOAA 2010). The categories and associated peak wind speeds in the SSHWS are the same as they were in the Saffir-Simpson Hurricane Scale.

> **CROSS REFERENCE**
>
> See Chapter 2 for a summary of the storms listed in Table 3-1. More details can be found in the "Coastal Flood and Wind Event Summaries" resource on the FEMA Residential Coastal Construction Web page.

Typhoons are divided into two categories; those with sustained winds less than 150 mph are referred to as typhoons, while those with sustained winds equal to or greater than 150 mph are known as ***super typhoons.***

Tropical cyclone records for the period 1851 to 2009 show that approximately one in five named storms (tropical storms and hurricanes) in the North Atlantic basin make landfall as hurricanes along the Atlantic or Gulf of Mexico coast of the United States. Figure 3-5 shows the average percentages of landfalling hurricanes in the United States.

Tropical cyclone landfalls are not evenly distributed on a geographic basis. In fact, the incidence of landfalls varies greatly. Approximately 40 percent of all U.S. landfalling hurricanes directly hit Florida, and 83 percent of Category 4 and 5 hurricane strikes have directly hit either Florida or Texas. Table 3-2 shows direct hurricane hits to the mainland U.S. from 1851 to 2009 categorized using the Saffir-Simpson Hurricane Scale.

Table 3-1. Saffir-Simpson Hurricane Wind Scale

Scale Number (Category)	Over Water Wind Speed in mph 1-Minute Sustained (3-Second Gust)	Property Damage	Examples[a]
1	74–95 (89–116)	Minimal	**Agnes** (1972 – Florida) **Earl** (1998 – Florida) **Dolly** (2008 – Texas)
2	96–110 (117–134)	Moderate	**Bob** (1991 – Rhode Island) **Marilyn** (1995 – U.S. Virgin Islands) **Frances** (2004 – Florida) **Ike** (2008 – Texas, Louisiana)
3	111–130 (135–159)	Extensive	**Alicia** (1983 – Texas) **Ivan** (2004 – Alabama)
4	131–155 (160–189)	Extreme	**Hugo** (1989 – South Carolina) **Andrew** (1992 – Florida) **Katrina** (2005 – Louisiana)
5	>155 (>189)	Catastrophic	**Florida Keys** (1935) **Camille** (1969 – Louisiana, Mississippi)

DATA SOURCE: NOAA HISTORICAL HURRICANE TRACKS (http://csc.noaa.gov/hurricanes)

(a) Hurricanes are listed according to their respective category at landfall based on wind speed.

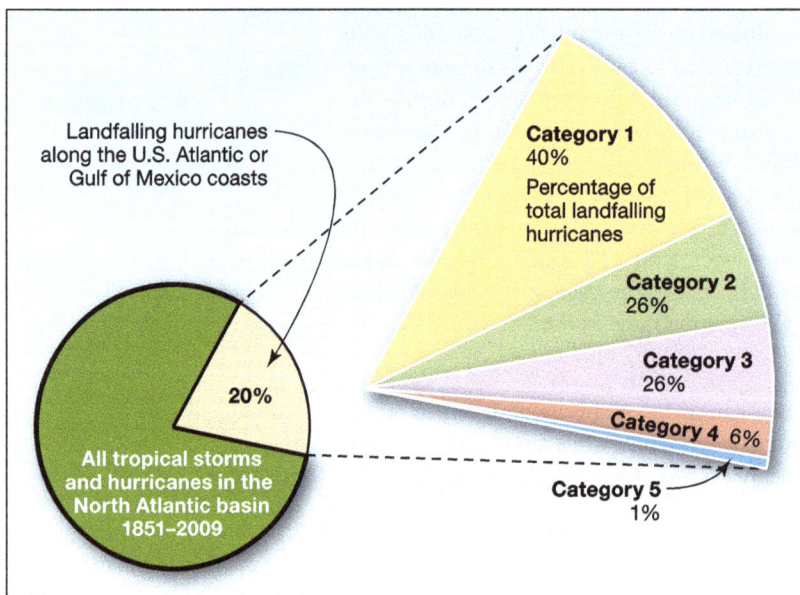

Figure 3-5. Classification (by Saffir-Simpson Hurricane scale) of landfalling tropical cyclones along the U.S. Atlantic and Gulf of Mexico coasts, 1851–2009

DATA SOURCES: BLAKE ET AL. 2005, JARRELL ET AL. 2001, NOAA 2011a

Table 3-2. Direct Hurricane Hits to U.S. Coastline Between 1851 and 2009 from Texas to Maine

Area	Saffir-Simpson Hurricane Scale Category					
	1	2	3	4	5	All
Texas	25	19	12	7	0	63
Louisiana	18	15	15	4	1	53
Mississippi	2	5	8	0	1	16
Alabama	12	5	6	0	0	23
Florida	44	33	29	6	2	114
Georgia	12	5	2	1	0	20
South Carolina	19	6	4	2	0	31
North Carolina	22	13	11	1	0	46
Virginia	9	2	1	0	0	12
Maryland	1	1	0	0	0	2
Delaware	2	0	0	0	0	2
New Jersey	2	0	0	0	0	2
Pennsylvania	1	0	0	0	0	1
New York	6	1	5	0	0	12
Connecticut	4	3	3	0	0	10
Rhode Island	3	2	4	0	0	9
Massachusetts	5	2	3	0	0	10
New Hampshire	1	1	0	0	0	2
Maine	5	1	0	0	0	6
Atlantic/Gulf U.S. Coastline (Texas to Maine)	**115**	**76**	**76**	**18**	**3**	**288**

DATA SOURCES: BLAKE ET AL. 2005, JARRELL ET AL. 2001, NOAA 2011a

Note: A direct hurricane hit means experiencing the core of strong winds and/or storm surge of a hurricane. State totals will not add up to U.S. totals because some storms are counted for more than one State

Another method of analyzing tropical cyclone incidence data is to compute the ***mean return period***, or the average time (in years) between landfall or nearby passage of a tropical storm or hurricane. Note that over short periods of time, the actual number and timing of tropical cyclone passage/landfall may deviate substantially from the long-term statistics. Some years see little tropical cyclone activity with no landfalling storms; other years see many storms with several landfalls. A given area may not experience the effects of a tropical cyclone for years or decades, and then be affected by several storms in a single year.

3.2.1.2   Other Coastal Storms

Other coastal storms include storms lacking closed circulation, but capable of producing strong winds. These storms usually occur during winter months and can affect the Atlantic coast, Pacific coast, the Great Lakes coast, and, rarely, the Gulf of Mexico coast. Along the ***Atlantic coast***, these storms are known as extratropical storms or nor'easters. Two of the most powerful and damaging nor'easters on record are the March 5–7, 1962 storm (see Figure 3-6) and the October 28–November 3, 1991 storm.

Coastal storms along the ***Pacific coast*** of the United States are usually associated with the passage of weather fronts during the winter months. These storms produce little or no storm surge (generally 2 feet or less) along the ocean shoreline, but they are capable of generating hurricane-force winds and large, damaging waves. Storm characteristics and patterns along the Pacific coast are strongly influenced by the occurrence of the El Niño Southern Oscillation (ENSO)—a climatic anomaly resulting in above-normal ocean temperatures and elevated sea levels along the U.S. Pacific coast. During El Niño years, sea levels along the Pacific shoreline tend to rise as much as 12 to 18 inches above normal, the incidence of coastal storms increases, and the typical storm track shifts from the Pacific Northwest to southern and central California. The net result of these effects is increased storm-induced erosion, changes in longshore sediment transport (due to changes in the direction of wave approach, which changes erosion/deposition patterns along the shoreline), and increases the incidence of rainfall and landslides in coastal regions.

Storms on the ***Great Lakes*** are usually associated with the passage of low-pressure systems or cold fronts. Storm effects (high winds, storm surge, and wave runup) may last a few hours or a few days. Storm surges and damaging wave conditions on the Great Lakes are a function of wind speed, direction, duration, and fetch; if high winds occur over a long fetch for more than an hour or so, the potential for flooding and erosion exists. However, because of the sizes and depths of the Great Lakes, storm surges are usually limited to less than 2 feet, except in embayments (2 to 4 feet) and on Lake Erie (up to 8 feet). Periods of active erosion are triggered by heavy precipitation events, storm waves, rising lake levels, and changes in groundwater outflow along the coast.

Figure 3-6.
Flooding, erosion, and overwash at Fenwick Island, DE, following March 1962 nor'easter

3.3   Coastal Hazards

This section addresses coastal hazards of high wind, earthquakes, tsunamis, and other hazards and environmental effects. Coastal flooding and erosion hazards are discussed separately, in Sections 3.4 and 3.5, respectively.

3.3.1   High Winds

High winds can originate from a number of events. Tropical storms, hurricanes, typhoons, other coastal storms, and tornadoes generate the most significant coastal wind hazards.

The most current design wind speeds are given by the national load standard, ASCE 7-10, *Minimum Design Loads for Buildings and Other Structures* (ASCE 2010). Figure 3-7, taken from ASCE 7-10, shows the geographic distribution of design wind speeds for the continental United States and Alaska, and lists design wind speeds for Hawaii, Puerto Rico, Guam, American Samoa, and the Virgin Islands. The Hawaii State Building Code includes detailed design wind speed maps for all four counties in Hawaii. They are available online at http://hawaii.gov/dags/bcc/comments/wind-maps-for-state-building-code.

High winds are capable of imposing large lateral (horizontal) and uplift (vertical) forces on buildings. Residential buildings can suffer extensive wind damage when they are improperly designed and constructed and when wind speeds exceed design levels (see Figures 3-8 and 3-9). The effects of high winds on a building depend on many factors, including:

- Wind speed (sustained and gusts) and duration of high winds
- Height of building above ground
- Exposure or shielding of the building (by topography, vegetation, or other buildings) relative to wind direction
- Strength of the structural frame, connections, and envelope (walls and roof)
- Shape of building and building components
- Number, size, location, and strength of openings (e.g., windows, doors, vents)
- Presence and strength of shutters or opening protection
- Type, quantity, and velocity of wind-borne debris

Even when wind speeds do not exceed design levels, such as during Hurricane Ike, residential buildings can suffer extensive wind damage when they are improperly designed and constructed. The beach house shown in Figure 3-10 experienced damage to its roof structure. The apartment building in Figure 3-11 experienced

> **NOTE**
>
> Basic wind speeds given by ASCE 7-10, shown in Figure 3-7 of this Manual, correspond to a wind with a recurrence interval of 700 years for Risk Category II buildings.
>
> The 2012 IRC contains a simplified table based on ASCE 7-10, which can be used to obtain an effective basic wind speed for sites where topographic wind effects are a concern.

> **NOTE**
>
> It is generally beyond the scope of most building designs to account for a direct strike by a tornado (the ASCE 7-10 wind map in Figure 3-7 excludes tornado effects). However, use of wind-resistant design techniques will reduce damage caused by a tornado passing nearby.
>
> Section 3.3.1.3 discusses tornado effects.

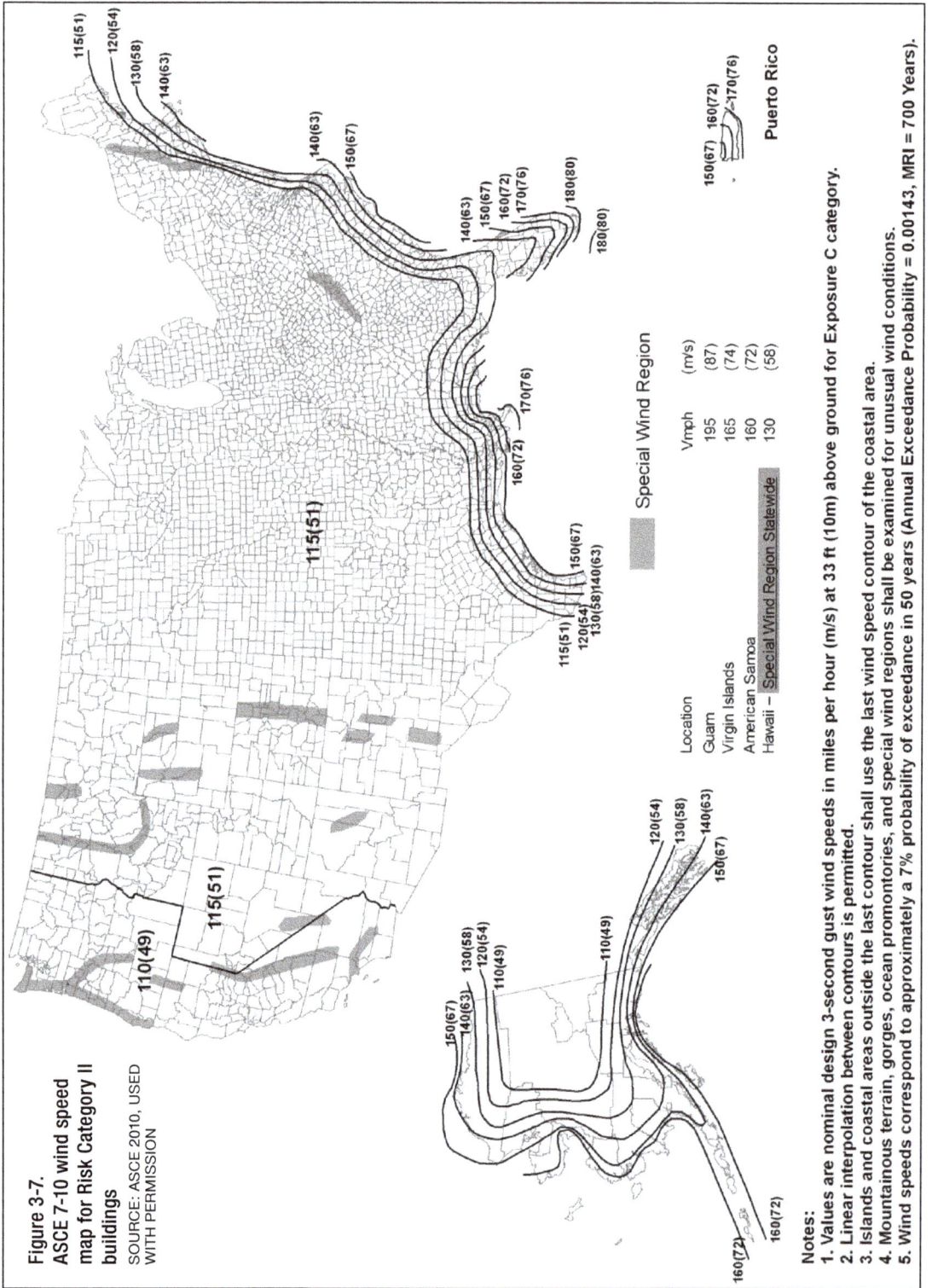

Figure 3-7.
ASCE 7-10 wind speed map for Risk Category II buildings
SOURCE: ASCE 2010, USED WITH PERMISSION

115(51)
120(54)
130(58)
140(63)
140(63)
150(67)
140(63)
150(67)
160(72)
170(76)
180(80)
180(80)
150(67) 160(72)
>170(76)
Puerto Rico

170(76)
160(72)
150(67)
140(63)
130(58)
120(54)
115(51)

115(51)

115(51)

110(49)

120(54)
130(58)
140(63)
150(67)

150(67)
140(63)
130(58)
120(54)
110(49)
110(49)

160(72)

160(72)

Special Wind Region

Location	Vmph	(m/s)
Guam	195	(87)
Virgin Islands	165	(74)
American Samoa	160	(72)
Hawaii – Special Wind Region Statewide	130	(58)

Notes:
1. Values are nominal design 3-second gust wind speeds in miles per hour (m/s) at 33 ft (10m) above ground for Exposure C category.
2. Linear interpolation between contours is permitted.
3. Islands and coastal areas outside the last contour shall use the last wind speed contour of the coastal area.
4. Mountainous terrain, gorges, ocean promontories, and special wind regions shall be examined for unusual wind conditions.
5. Wind speeds correspond to approximately a 7% probability of exceedance in 50 years (Annual Exceedance Probability = 0.00143, MRI = 700 Years).

Figure 3-8.
End-wall failure of typical first-floor masonry/second-floor wood-frame building in Dade County, FL (Hurricane Andrew, 1992)

Figure 3-9.
Loss of roof sheathing due to improper nailing design and schedule in Kauai County, HI (Hurricane Iniki, 1992)

Figure 3-10.
Beach house with roof structure removed by Hurricane Ike (Galveston, TX, 2008)

Figure 3-11.
Apartment building with gable end wind damage from Hurricane Ike as a result of poor connection between brick veneer and wall structure (Galveston, TX, 2008)

gable end wall damage when the wall sheathing failed as a result of a poor connection between the brick veneer and the stud walls.

Proper design and construction of residential structures, particularly those close to open water or near the coast, demand that every factor mentioned above be investigated and addressed carefully. Failure to do so may ultimately result in building damage or destruction by wind.

Three wind-related topics that deserve special attention from design professionals are speedup of wind due to topographic effects, wind-borne debris and rainfall penetration into buildings, and tornadoes.

3.3.1.1   Speedup of Winds Due to Topographic Effects

Speedup of winds due to topographic effects can occur wherever mountainous areas, gorges, and ocean promontories exist. Thus, the potential for increased wind speeds should be investigated for any construction on or near the crests of high coastal bluffs, cliffs, or dunes, or in gorges and canyons. ASCE 7-10 provides guidance on calculating increased wind speeds in such situations.

Designers should also consider the effects of long-term erosion on the wind speeds a building may experience over its lifetime. For example, a building sited atop a tall bluff, but away from the bluff edge, is not prone to wind speedup initially, but long-term erosion may move the bluff edge closer to the building and expose the building to increased wind speeds due to topographic changes.

3.3.1.2   Wind-Borne Debris and Rainfall Penetration

Wind loads and wind-borne debris are both capable of causing damage to a building envelope. Even small failures in the building envelope, at best, lead to interior damage by rainfall penetration and winds and, at worst, lead to internal pressurization of the building, roof loss, and complete structural disintegration.

Sparks et al. (1994) investigated the dollar value of insured wind losses following Hurricanes Hugo and Andrew and found the following:

COST CONSIDERATION

Even minor damage to the building envelope can lead to large economic losses, as the building interior and contents get wet.

- Most wind damage to houses is restricted to the building envelope

- Rainfall entering a building through envelope failures causes the dollar value of direct building damage to be magnified by a factor of two (at lower wind speeds) to nine (at higher wind speeds)

- Lower levels of damage magnification are associated with water seeping through exposed roof sheathing (e.g., following loss of shingles or roof tiles)

- Higher levels of damage magnification are associated with rain pouring through areas of lost roof sheathing and through broken windows and doors

3.3.1.3 Tornadoes

A tornado is a rapidly rotating vortex or funnel of air extending groundward from a cumulonimbus cloud. Tornadoes are spawned by severe thunderstorms and by hurricanes. Tornadoes often form in the right forward quadrant of a hurricane, far from the hurricane eye. The strength and number of tornadoes are not related to the strength of the hurricane that generates them. In fact, the weakest hurricanes often produce the most tornadoes. Tornadoes can lift and move huge objects, move or destroy houses, and siphon large volumes from bodies of water. Tornadoes also generate large amounts of debris, which then become wind-borne and cause additional damage.

CROSS REFERENCE

The FEMA MAT program has published several MAT reports and recovery advisories following tornado disasters in the United States. These publications offer both insight into the performance of buildings during tornadoes and solutions. To obtain copies of these publications, see the FEMA MAT Web page (http://www.fema.gov/rebuild/mat).

Tornadoes are rated using the Enhanced Fujita (EF) Scale, which correlates tornado wind speeds to categories EF0 through EF5 based on damage indicators and degrees of damage. Table 3-3 shows the EF Scale. For more information on how to assess tornado damage based on the EF Scale, refer to *A Recommendation for an Enhanced Fujita Scale* by the Texas Tech Wind Science and Engineering Center at http://www.spc.noaa.gov/faq/tornado/ef-ttu.pdf (TTU 2004).

Table 3-3. Enhanced Fujita Scale in Use Since 2007

EF Scale Rating	3-Second Gust Speed (mph)	Type of Damage
EF0	65–85	Light damage
EF1	86–110	Moderate damage
EF2	111–135	Considerable damage
EF3	136–165	Severe damage
EF4	166–200	Devastating damage
EF5	>200	Incredible damage

Hardened buildings and newer structures designed and constructed to modern, hazard-resistant codes can generally resist the wind loads from weak tornadoes. When stronger tornadoes strike, not all damage is from the rotating vortex of the tornado. Much of the damage is caused by straight-line winds being pulled into and rushing toward the tornado itself. Homes built to modern codes may survive some tornadoes without structural failure, but often experience damage to the cladding, roof covering, roof deck, exterior walls, and windows. For most building uses, it is economically impractical to design the entire building to resist tornadoes. Portions of buildings can be designed as safe rooms to protect occupants from tornadoes.

> **CROSS REFERENCE**
>
> FEMA 320, *Taking Shelter from the Storm: Building a Safe Room for Your Home or Small Business* (FEMA 2008a) provides guidance and designs for residential safe rooms that provide near-absolute protection against the forces of extreme winds. For more information, see the FEMA safe room Web page (http://www.fema.gov/plan/prevent/saferoom/index.shtm).

3.3.2   Earthquakes

Earthquakes can affect coastal areas just as they can affect inland areas through ground shaking, liquefaction, surface fault ruptures, and other ground failures. Therefore, coastal construction in seismic hazard areas must take potential earthquake hazards into account. Since basic principles of earthquake-resistant design can contradict flood-resistant design principles, proper design in coastal seismic hazard areas must strike a balance between:

- The need to elevate buildings above flood hazards and minimize obstructions to flow and waves beneath a structure

- The need to stabilize or brace the building against potentially violent accelerations and shaking due to earthquakes

Earthquakes are classified according to magnitude and intensity. Magnitude refers to the total energy released by the event. Intensity refers to the effects at a particular site. Thus, an earthquake has a single magnitude, but the intensity varies with location. The Richter Scale is used to report earthquake magnitude, while the Modified Mercalli Intensity (MMI) Scale is used to report felt intensity. The MMI Scale (see Table 3-4) ranges from I (imperceptible) to XII (catastrophic).

> **CROSS REFERENCE**
>
> Seismic load provisions and earthquake ground motion maps can be found in the following codes and standards:
>
> - IBC Section 1613
> - IRC R301.2.2
> - ASCE 7 Chapters 11 through 23
>
> For best practices guidance, see FEMA 232, *Homebuilders' Guide to Earthquake Resistant Design and Construction* (FEMA 2006a).

The ground motion produced by earthquakes can shake buildings (laterally and vertically) and cause structural failure by excessive deflection. Earthquakes can cause building failures by rapid uplift, subsidence, ground rupture, soil liquefaction, or consolidation. In coastal areas, the structural effects of ground shaking can be magnified when buildings are elevated above the natural ground elevation to mitigate flooding.

One of the site parameters controlling seismic-resistant design of buildings is the maximum considered earthquake ground motion, which is defined in the IBC as the most severe earthquake effects considered in the IBC, and has been mapped based on the 0.2-second spectral response acceleration and the 1.0-second spectral response acceleration as a percent of the gravitational constant ("g").

Table 3-4. Earthquake MMI Scale

MMI Level	Felt Intensity
I	Not felt except by very few people under special conditions. Detected mostly by instruments.
II	Felt by a few people, especially those on the upper floors of buildings. Suspended objects may swing.
III	Felt noticeably indoors. Standing automobiles may rock slightly.
IV	Felt noticeably indoors, by a few outdoors. At night, some people may be awakened. Dishes, windows, and doors rattle.
V	Felt by nearly everyone. Many people are awakened. Some dishes and windows are broken. Unstable objects are overturned.
VI	Felt by nearly everyone. Many people become frightened and run outdoors. Some heavy furniture is moved. Some plaster falls.
VII	Most people are alarmed and run outside. Damage is negligible in buildings of good construction, considerable in buildings of poor construction.
VIII	Damage is slight in specially designed structures, considerable in ordinary buildings, great in poorly built structures. Heavy furniture is overturned.
IX	Damage is considerable in specially designed buildings. Buildings shift from their foundations and partly collapse. Underground pipes are broken.
X	Some well-built wooden structures are destroyed. Most masonry structures are destroyed. The ground is badly cracked. Considerable landslides occur on steep slopes.
XI	Few, if any, masonry structures remain standing. Rails are bent. Broad fissures appear in the ground.
XII	Virtually total destruction. Waves are seen on the ground surface. Objects are thrown in the air.

SOURCE: FEMA 1997

The structural effects of earthquakes are a function of many factors (e.g., soil characteristics; local geology; and building weight, shape, height, structural system, and foundation type). Design of earthquake-resistant buildings requires careful consideration of both site and structure.

In many cases, elevating a building 8 to 10 feet above grade on a pile or column foundation—a common practice in low-lying Zone V and Coastal A Zone areas—can result in what earthquake engineers term an "inverted pendulum" as well as a discontinuity in the floor diaphragm and vertical lateral force-resisting system. Both conditions require the building be designed for a larger earthquake force. Thus, designs for pile- or column-supported residential buildings should be verified for necessary strength and rigidity below the first-floor level (see Chapter 10) to account for increased stresses in the foundation members during an earthquake. For buildings elevated on fill, earthquake ground motions can be exacerbated if the fill and underlying soils are not properly compacted and stabilized.

Liquefaction of the supporting soil can be another damaging consequence of ground shaking. In granular soils with high water tables (like those found in many coastal areas), the ground motion can create a semi-liquid soil state. The soil then can temporarily lose its bearing capacity, and settlement and differential movement of buildings can result.

Seismic effects on buildings vary with structural configuration, stiffness, ductility, and strength. Properly designed and built wood-frame buildings are quite ductile, meaning that they can withstand large deformations without losing strength. Failures, when they occur in wood-frame buildings, are usually at connections. Properly designed and built steel construction is also inherently ductile, but can fail at non-ductile connections, especially at welded connections. Bolted connections have performed better than welded connections under seismic loads. Modern concrete construction can be dimensioned and reinforced to provide sufficient strength and ductility to resist earthquakes; older concrete structures are typically more vulnerable. Elements of existing concrete structures can be retrofitted with a variety of carbon-fiber, glass-fiber, glass-fiber-reinforced or fiber-reinforced polymer wraps and strips to increase the building's resistance to seismic effects, although this is typically a costly option. Failures in concrete masonry structures are likely to occur if reinforcing and cell grouting do not meet seismic-resistant requirements.

3.3.3   Tsunamis

Tsunamis are long-period water waves generated by undersea shallow-focus earthquakes, undersea crustal displacements (subduction of tectonic plates), landslides, or volcanic activity. Tsunamis can travel great distances, undetected in deep water, but shoaling rapidly in coastal waters and producing a series of large waves capable of destroying harbor facilities, shore protection structures, and upland buildings (see Figure 3-12). Tsunamis have been known to damage some structures thousands of feet inland and over 50 feet above sea level.

Coastal construction in tsunami hazard zones must consider the effects of tsunami runup, flooding, erosion, and debris loads. Designers should also be aware that the "rundown" or return of water to the sea can also damage the landward sides of structures that withstood the initial runup.

> **NOTE**
>
> Information about tsunamis and their effects is available from the National Tsunami Hazard Mitigation Program Web site: http:// nthmp.tsunami.gov.

Tsunami effects at a site are determined by four basic factors:

- Magnitude of the earthquake or triggering event
- Location of the triggering event
- Configuration of the continental shelf and shoreline
- Upland topography

Figure 3-12. Damage from the 2009 tsunami (Amanave, American Samoa)

SOURCE: ASCE, USED WITH PERMISSION

The ***magnitude of the triggering event*** determines the period of the resulting waves, and generally (but not always) the tsunami magnitude and damage potential. Unlike typical wind-generated water waves with periods between 5 and 20 seconds, tsunamis can have wave periods ranging from a few minutes to over 1 hour (Camfield 1980). As wave periods increase, the potential for coastal inundation and damage also increases. Wave period is also important because of the potential for resonance and wave amplification within bays, harbors, estuaries, and other semi-enclosed bodies of coastal water.

The ***location of the triggering event*** has two important consequences. First, the distance between the point of tsunami generation and the shoreline determines the maximum available warning time. Tsunamis generated at a remote source take longer to reach a given shoreline than locally generated tsunamis.

Second, the point of generation determines the direction from which a tsunami approaches a given site. Direction of approach can affect tsunami characteristics at the shoreline because of the sheltering or amplification effects of other land masses and offshore bathymetry. The ***configuration of the continental shelf and shoreline*** affect tsunami impacts at the shoreline through wave reflection, refraction, and shoaling. Variations in offshore bathymetry and shoreline irregularities can focus or disperse tsunami wave energy along certain shoreline reaches, increasing or decreasing tsunami impacts.

Upland elevations and topography also determine tsunami impacts at a site. Low-lying tsunami-prone coastal sites are more susceptible to inundation, tsunami runup, and damage than sites at higher elevations.

Table 3-5 lists areas where tsunami events have been observed in the United States and its territories, and the sources of those events. Note that other areas may be subject to rare tsunami events.

Table 3-5. Areas of Observed Tsunami Events in the United States and Territories

Area		Principal Source of Tsunamis
Alaska:	North Pacific coast	Locally generated events (landslides, subduction, submarine landslides, volcanic activity)
	Aleutian Islands	Locally generated events and remote source earthquakes
	Gulf of Alaska coast	Locally generated events and remote source earthquakes
Hawaii		Locally generated events and remote source earthquakes
American Samoa		Locally generated events and remote source earthquakes
Oregon		Locally generated events and remote source earthquakes
Washington		Locally generated events and remote source earthquakes
California		Locally generated events and remote source earthquakes
Puerto Rico		Locally generated events
U.S. Virgin Islands		Locally generated events

3.3.4 Other Hazards and Environmental Effects

Other hazards to which coastal construction may be exposed include a wide variety of hazards whose incidence and severity may be highly variable and localized. Examples include subsidence and uplift, landslides and ground failures, salt spray and moisture, rain, hail, wood decay and termites, wildfires, floating ice, snow, and atmospheric ice. These hazards do not always come to mind when coastal hazards are mentioned, but like the other hazards described in this chapter, they can affect coastal construction and should be considered in siting, design, and construction decisions.

3.3.4.1   Sea and Lake Level Rise

Coastal flood effects, described in detail in Section 3.4, typically occur over a period of hours or days. However, longer-term water level changes also occur. Sea level tends to rise or fall over centuries or thousands of years, in response to long-term global climate changes. Great Lakes water levels fluctuate both seasonally and over decades in response to regional climate changes. In either case, medium- and long-term increases in water levels increase the damage-causing potential of coastal flood and storm events and often cause a permanent horizontal recession of the shoreline.

Global mean sea level has been rising at long-term rates averaging 1.7 (+/-0.5) millimeters annually for the twentieth century (over 6 inches total during the twentieth century) (Intergovernmental Panel on Climate Change [IPCC] 2007). Rates of mean sea level rise along the Louisiana and Texas coasts, as well as portions of the Atlantic coast, are significantly higher than the global average (as high as 3.03 feet per century in Grand Isle, LA). Records for U.S. Pacific coast stations show that some areas have experienced rises in relative sea levels of over 1 foot per century. Other areas have experienced a fall in relative sea levels; Alaska's relative sea level fall rate is as high as 3.42 feet per century (see Figure 3-13).

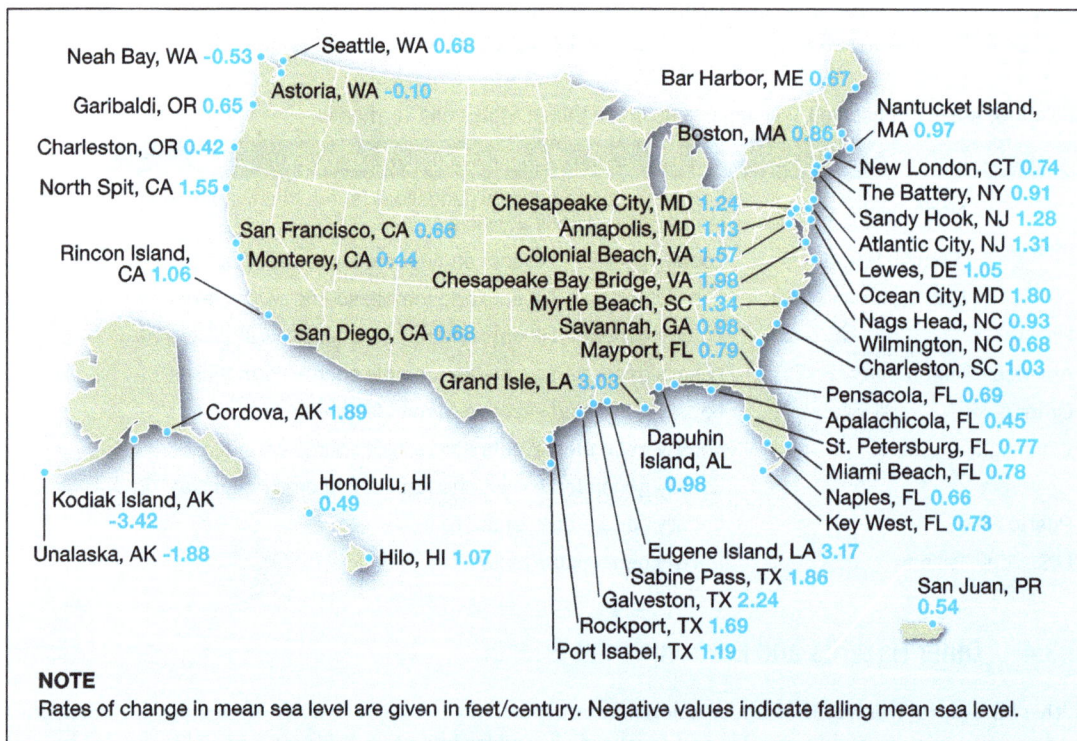

NOTE
Rates of change in mean sea level are given in feet/century. Negative values indicate falling mean sea level.

Figure 3-13.
Observations of rates of change in mean sea level in the United States in feet per century
DATA SOURCE: NOAA CENTER FOR OPERATIONAL OCEANOGRAPHIC PRODUCTS AND SERVICES
(http://tidesandcurrents.noaa.gov/sltrends/sltrends.html)

Detailed historical and recent sea level data for U.S. coastal stations are available from NOAA Center for Operational Oceanographic Products and Services at http://tidesandcurrents.noaa.gov/sltrends/sltrends.html (see Figure 3-14 for an example of mean sea level trend for a station in Atlantic City, NJ).

The EPA provides links to recent reports (including those of the IPCC) and data at http://www.epa.gov/climatechange/science/recentslc.html.

CROSS REFERENCE

For more information on measured and projected Great Lakes water levels, see the USACE Detroit District Monthly Bulletin of Great Lakes Water Levels Web page at http://www.lre.usace.army.mil/greatlakes/hh/greatlakeswaterlevels/waterlevelforecasts/monthlybulletinofgreatlakeswaterlevels.

Great Lakes water-level records dating from 1860 are maintained by the USACE Detroit District. The records show seasonal water levels typically fluctuate between 1 and 2 feet. The records also show that long-term (approximately 100 years) water levels in Lakes Michigan, Huron, Erie, and Ontario have fluctuated approximately 6 feet, and water levels in Lake Superior have fluctuated approximately 4 feet. Figure 3-15 shows a typical plot of actual and projected lake levels for Lakes Michigan and Huron.

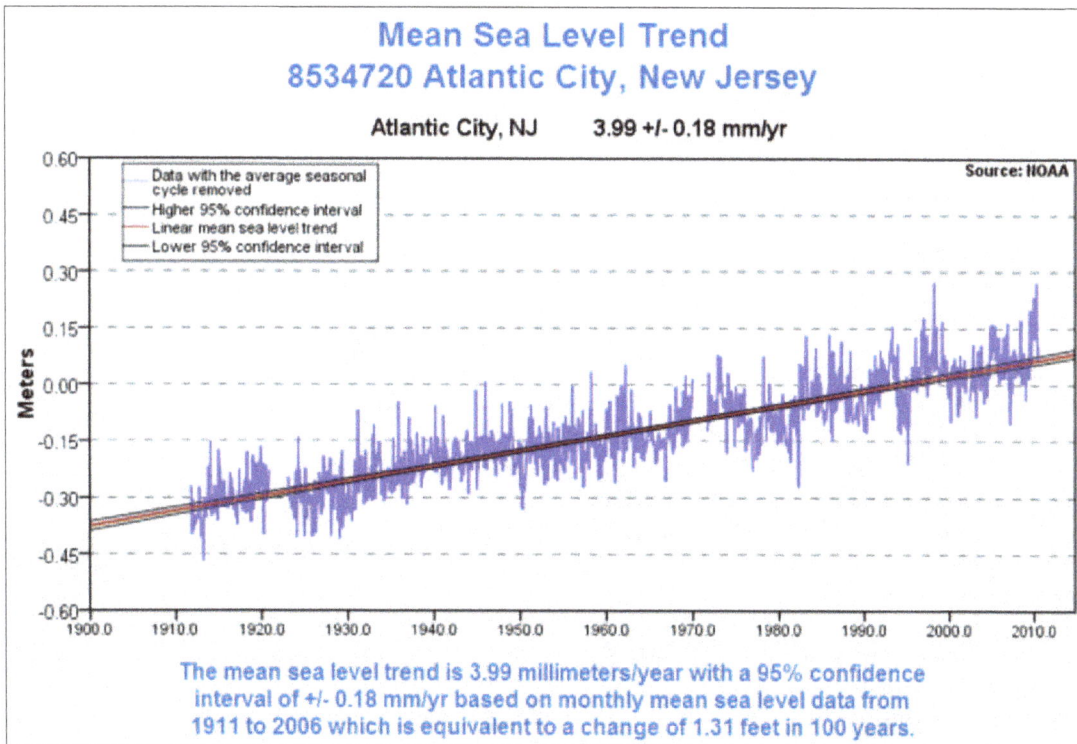

Figure 3-14.
Mean sea level rise data for a station in Atlantic City, NJ
SOURCE: NOAA 2011b

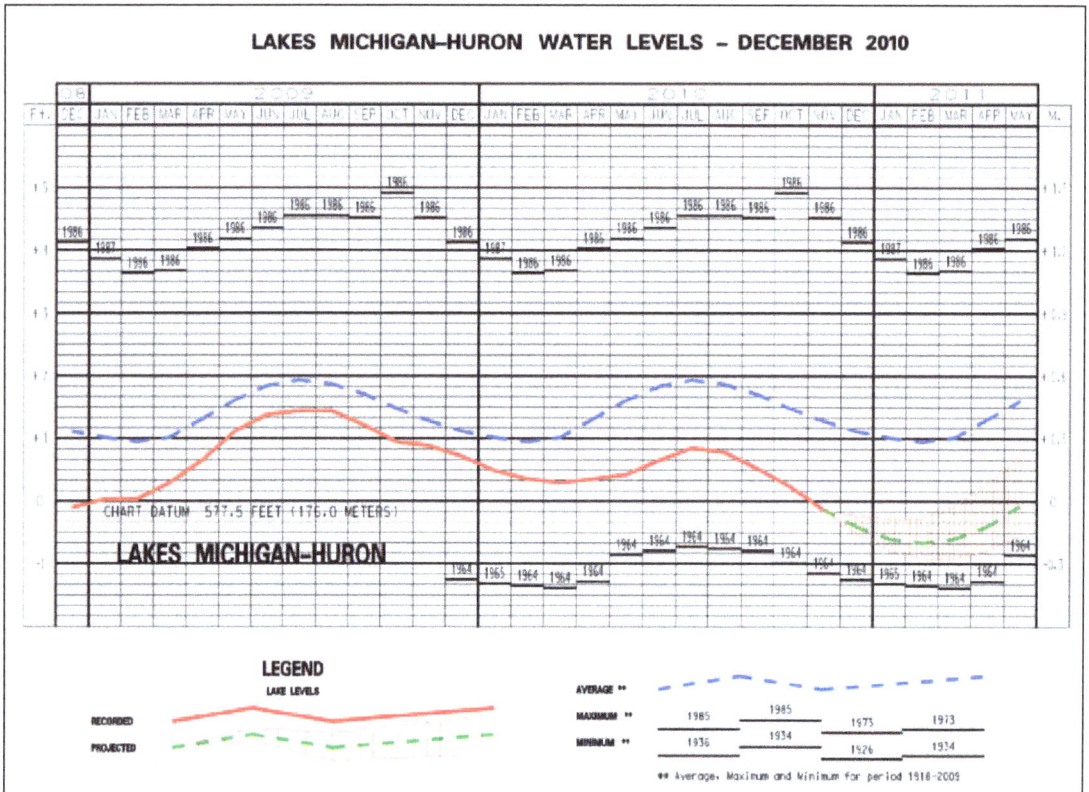

Figure 3-15.
Monthly bulletin of lake levels for Lakes Michigan and Huron
SOURCE: USACE DETROIT DISTRICT, ACCESSED DECEMBER 2010

Keillor (1998) discusses the implications of both high and low lake levels on Great Lakes shorelines. In general, beach and bluff erosion rates tend to increase as water levels rise over a period of several years, such as occurred in the mid-1980s. As water levels fall, erosion rates diminish. Low lake levels lead to generally stable shorelines and bluffs, but make navigation through harbor entrances difficult (see Section 3.5 for more information on coastal bluff erosion).

Designers, community officials, and owners should note that FIRMs do not account for sea level rise or Great Lakes water level trends. Relying on FIRMs for estimates of elevations for future water and wave effects is not advised for any medium- to long-term planning horizon (10 to 20 years or longer). Instead, forecasts of future water levels should be incorporated into project planning. This has been done at the Federal level in the USACE publication titled *Water Resource Policies and Authorities Incorporating Sea-Level Change Considerations in Civil Works Program* (USACE 2009a), which includes guidance on where to obtain water level change information and how to interpret and use such information. The USACE publication contains a flow chart

NOTE

Because coastal land masses can move up (uplift) or down (subsidence) independent of water levels, discussions related to water level change must be expressed in terms of relative sea level or relative lake level.

and a step-by-step process to follow. Although the publication was written with USACE projects in mind, the guidance will be helpful to those planning and designing coastal residential buildings.

3.3.4.2   Subsidence and Uplift

Subsidence is a hazard that typically affects areas where (1) withdrawal of groundwater or petroleum has occurred on a large scale, (2) organic soils are drained and settlement results, (3) younger sediments deposit over older sediments and cause those older sediments to compact (e.g., river delta areas), or (4) surface sediments collapse into underground voids. The last of these four is most commonly associated with mining and rarely affects coastal areas (coastal limestone substrates would be an exception because these areas could be affected by collapse). The remaining three causes (groundwater or petroleum withdrawal, organic soil drainage, and sediment compaction) have all affected coastal areas in the past (FEMA 1997). One consequence of coastal subsidence, even when small in magnitude, is an increase in coastal flood hazards due to an increase in flood depth. For example, Figure 3-16 shows land subsidence in the Houston-Galveston area. In portions of Texas, subsidence has been measured for over 100 years, and subsidence of several feet has been recorded over a wide area; some land areas in Texas have dropped 10 feet in elevation since 1906. Subsidence also complicates flood hazard mapping and can render some flood hazard maps obsolete before they would otherwise need to be updated.

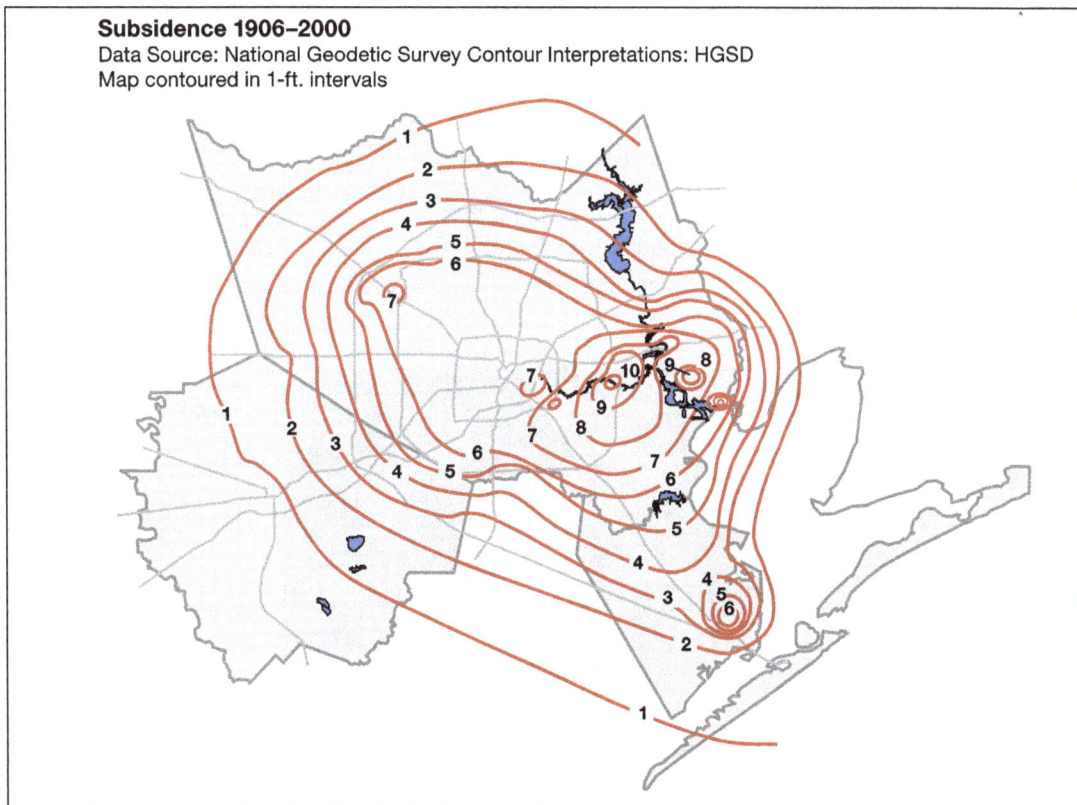

Subsidence 1906–2000
Data Source: National Geodetic Survey Contour Interpretations: HGSD
Map contoured in 1-ft. intervals

Figure 3-16.
Land subsidence in the Houston-Galveston area, 1906–2000
SOURCE: HARRIS-GALVESTON SUBSIDENCE DISTRICT 2010

Land uplift is the result of the ground rising due to various geological processes. Although few people regard land uplift as a coastal hazard, Larsen (1994) has shown that differential uplift in the vicinity of the Great Lakes can lead to increased water levels and flooding. As the ground rises in response to the removal of the great ice sheet, it does so in a non-uniform fashion. On Lake Superior, the outlet at the eastern end of the lake is rising at a rate of nearly 10 inches per century, relative to the city of Duluth-Superior at the western end of the lake. This causes a corresponding water level rise at Duluth-Superior. Similarly, the northern ends of Lakes Michigan and Huron are rising relative to their southern portions. On Lake Michigan, the northern outlet at the Straits of Mackinac is rising at a rate of 9 inches per century, relative to Chicago, at the southern end of the lake. The outlet of Lakes Michigan and Huron is rising only about 3 inches per century relative to the land at Chicago.

3.3.4.3   Salt Spray and Moisture

Salt spray and moisture effects frequently lead to corrosion and decay of building materials in the coastal environment. These hazards are commonly overlooked or underestimated by designers. Any careful inspection of coastal buildings (even new or recent buildings) near a large body of water will reveal deterioration of improperly selected or installed materials.

For example, metal connectors, straps, and clips used to improve a building's resistance to high winds and earthquakes often show signs of corrosion (see Figure 3-17). Corrosion is affected by many factors, but the primary difference between coastal and inland/Great Lakes areas is the presence of salt spray, tossed into the air by breaking waves and blown onto land by onshore winds. Salt spray accumulates on metal surfaces, accelerating the electrochemical processes that cause corrosion, particularly in the humid conditions common along the coast.

Corrosion severity varies considerably from community to community along the coast, from building to building within a community, and even within an individual building.

> **CROSS REFERENCE**
>
> See Chapter 14, Section 14.2, for a discussion of salt spray and moisture effects.

Figure 3-17.
Example of corrosion, and resulting failure, of metal connectors
SOURCE: SPENCER ROGERS, USED WITH PERMISSION

Factors affecting the rate of corrosion include humidity, wind direction and speed, seasonal wave conditions, distance from the shoreline, elevation above the ground, orientation of the building to the shoreline, rinsing by rainfall, shelter and air flow in and around the building, and the component materials.

CROSS REFERENCE

See FEMA Technical Bulletin 8, *Corrosion Protection for Metal Connectors in Coastal Areas* (1996), for more information about corrosion and corrosion-resistant connectors.

Wood decay is most commonly caused by moisture. Moisture-related decay is prevalent in all coastal areas—it is not exclusive to buildings near the shoreline. Protection against moisture-related decay can be accomplished by one or more of the following: use of preservative-treated or naturally durable wood, proper detailing of wood joints to eliminate standing water, avoidance of cavity wall systems, and proper installation of water-resistive barriers. Sunlight, aging, insects, chemicals, and temperature can also lead to decay. FEMA P-499 Fact Sheet 1.7, *Coastal Building Materials*, has more information on the use of materials to resist corrosion, moisture, and decay (FEMA 2010).

3.3.4.4 Rain

Rain presents two principal hazards to coastal residential construction:

- Penetration of the building envelope during high-wind events (see Section 3.3.1.2)
- Vertical loads due to rainfall ponding on the roof

Ponding usually occurs on flat or low-slope roofs where a parapet or other building element causes rainfall to accumulate, and where the roof drainage system fails. Every inch of accumulated rainfall causes a downward-directed load of approximately 5 pounds per square foot. Excessive accumulation can lead to progressive deflection and instability of roof trusses and supports.

3.3.4.5 Hail

Hailstorms develop from severe thunderstorms, and generate balls or lumps of ice capable of damaging agricultural crops, buildings, and vehicles. Severe hailstorms can damage roofing shingles and tiles, metal roofs, roof sheathing, skylights, glazing, and other building components. Accumulation of hail on flat or low-slope roofs, like the accumulation of rainfall, can lead to significant vertical loads and progressive deflection of roof trusses and supports.

3.3.4.6 Termites

Infestation by termites is common in coastal areas subject to high humidity and frequent and heavy rains. Improper preservative treatments, improper design and construction, and even poor landscaping practices, can all contribute to infestation problems. The IRC includes a termite infestation probability map, which shows that most coastal areas have a moderate to very heavy probability of infestation (ICC 2012b).

Protection against termites can be accomplished by one or more of the following: use of preservative-treated wood products (including field treatment of notches, holes, and cut ends), use of naturally termite-resistant wood species, chemical soil treatment, and installation of physical barriers to termites (e.g., metal or plastic termite shields).

3.3.4.7   Wildfire

Wildfires can occur virtually everywhere in the United States and can threaten buildings constructed in coastal areas. Topography, the availability of vegetative fuel, and weather are the three principal factors that influence wildfire hazards. FEMA has produced several reports discussing the reduction of the wildfire hazard and the vulnerability of structures to wildfire hazards, including *Wildfire Mitigation in the 1998 Florida Wildfires* (FEMA 1998) and FEMA P-737, *Home Builder's Guide to Construction in Wildfire Zones* (FEMA 2008b). Some communities have adopted the *International Wildland-Urban Interface Code* (ICC 2012c), which includes provisions that address the spread of fire and defensible space for buildings constructed near wildland areas.

Experience with wildfires has shown that the use of fire-rated roof assemblies is one of the most effective methods of preventing loss of buildings to wildfire. Experience has also shown that replacing highly flammable vegetation around buildings with minimally flammable vegetation is also an effective way of reducing possible wildfire damage. Clearing vegetation around some buildings may be appropriate, but this action can lead to slope instability and landslide failures on steeply sloping land. Siting and construction on steep slopes requires careful consideration of multiple hazards with sometimes conflicting requirements.

3.3.4.8   Floating Ice

Some coastal areas of the United States are vulnerable to problems caused by floating ice. These problems can take the form of erosion and gouging of coastal shorelines, flooding due to ice jams, and lateral and vertical ice loads on shore protection structures and coastal buildings. On the other hand, the presence of floating ice along some shorelines reduces erosion from winter storms and wave effects. Designers should investigate potential adverse and beneficial effects of floating ice in the vicinity of their building site. Although this Manual does not discuss these issues in detail, additional information can be found in Caldwell and Crissman (1983), Chen and Leidersdorf (1988), and USACE (2002).

3.3.4.9   Snow

The principal hazard associated with snow is its accumulation on roofs and the subsequent deflection and potential failure of roof trusses and supports. Calculation of snow loads is more complicated than rain loads, because snow can drift and be distributed non-uniformly across a roof. Drainage of trapped and melted snow, like the drainage of rain water, must be addressed by the designer. In addition, particularly in northern climates such as New England and the Great Lakes, melting snow can result in ice dams. Ice dams can cause damage to roof coverings, drip edges, gutters, and other elements along eaves, leaving them more susceptible to future wind damage.

CROSS REFERENCE

Chapter 7 of ASCE 7 includes maps and equations for calculating snow loads. It also includes provisions for additional loads due to ice dams (ASCE 2010).

CROSS REFERENCE

State CZM programs (see Section 5.6, in Chapter 5) are a good source of hazard information, vulnerability analyses, mitigation plans, and other information about coastal hazards.

3.3.4.10   Atmospheric Ice

Ice can sometimes form on structures as a result of certain atmospheric conditions or processes (e.g., freezing rain or drizzle or in-cloud icing—accumulation of ice as supercooled clouds or fog comes into contact with a structure). The formation and accretion of this ice is termed **_atmospheric ice_**. Fortunately, typical coastal residential buildings are not considered ice-sensitive structures and are not subject to structural failures resulting from atmospheric ice. However, designers should consider proximity of coastal residential buildings to ice-sensitive structures (e.g., utility towers, utility lines, and similar structures) that may fail under atmospheric ice conditions. Designers should also be aware that ice build-up on structures, trees, and utility lines can result in a falling ice hazard to building occupants.

3.4   Coastal Flood Effects

Coastal flooding can originate from a number of sources. Tropical cyclones, other coastal storms, and tsunamis generate the most significant coastal flood hazards, which usually take the form of hydrostatic forces, hydrodynamic forces, wave effects, and flood-borne debris effects. Regardless of the source of coastal flooding, a number of flood parameters must be investigated at a coastal site to correctly characterize potential flood hazards:

- Origin of flooding
- Flood frequency
- Flood depth
- Flood velocity
- Flood direction

- Flood duration
- Wave effects
- Erosion and scour
- Sediment overwash
- Flood-borne debris

CROSS REFERENCE

See Section 8.5 for procedures used to calculate flood loads.

If a designer can determine each of these parameters for a site, the specification of design flood conditions is straightforward and the calculation of design flood loads will be more precise. Unfortunately, determining some of these parameters (e.g., flood velocity, debris loads) is difficult for most sites, and design flood conditions and loads may be less exact.

3.4.1   Hydrostatic Forces

Standing water or slowly moving water can induce horizontal hydrostatic forces against a structure, especially when floodwater levels on different sides of the structure are not equal. Also, flooding can cause vertical hydrostatic forces, or flotation (see Figure 3-18).

3.4.2   Hydrodynamic Forces

Hydrodynamic forces on buildings are created when coastal floodwaters move at high velocities. These high-velocity flows are capable of destroying solid walls and dislodging buildings with inadequate foundations. High-velocity flows can also move large quantities of sediment and debris that can cause additional damage.

High-velocity flows in coastal areas are usually associated with one or more of the following:

- Storm surge and wave runup flowing landward, through breaks in sand dunes or across low-lying areas (see Figure 3-19)

CROSS REFERENCE

Predicting the speed and direction of high-velocity flows is difficult. Designers should refer to the guidance contained in Section 8.5.6 and should assume that the flow can originate from any direction.

Figure 3-18.
Intact houses floated off their foundations and carried inland during Hurricane Hugo in 1989 (Garden City, SC)

Figure 3-19.
Storm surge at Horseshoe Beach, FL, during Tropical Storm Alberto in 2006
SOURCE: NOAA NATIONAL WEATHER SERVICE FORECAST OFFICE

- Tsunamis

- Outflow (flow in the seaward direction) of floodwaters driven into bay or upland areas

- Strong currents parallel to the shoreline, driven by the obliquely incident storm waves

NOTE

Storm surge does not correlate to hurricane category according to the earlier Saffir-Simpson Hurricane Scale, so the scale was renamed (Saffir Simpson Hurricane Wind Scale) and changed in 2010 to eliminate any reference to storm surge (see Table 3-1).

High-velocity flows can be created or exacerbated by the presence of manmade or natural obstructions along the shoreline and by weak points formed by shore-normal roads and access paths that cross dunes, bridges or shore-normal canals, channels, or drainage features. For example, evidence after Hurricane Opal struck Navarre Beach, FL, in 1995 suggests that large engineered buildings channeled flow between them (see Figure 3-20). The channelized flow caused deep scour channels across the island, undermining a pile-supported house between the large buildings (see Figure 3-21), and washing out roads and houses (see Figure 3-22) situated farther landward.

Figure 3-20.
Flow channeled between large buildings during Hurricane Opal in 1995 scoured a deep channel and damaged infrastructure and houses at Navarre Beach, FL

SOURCE: FLORIDA DEPARTMENT OF ENVIRONMENTAL PROTECTION, USED WITH PERMISSION

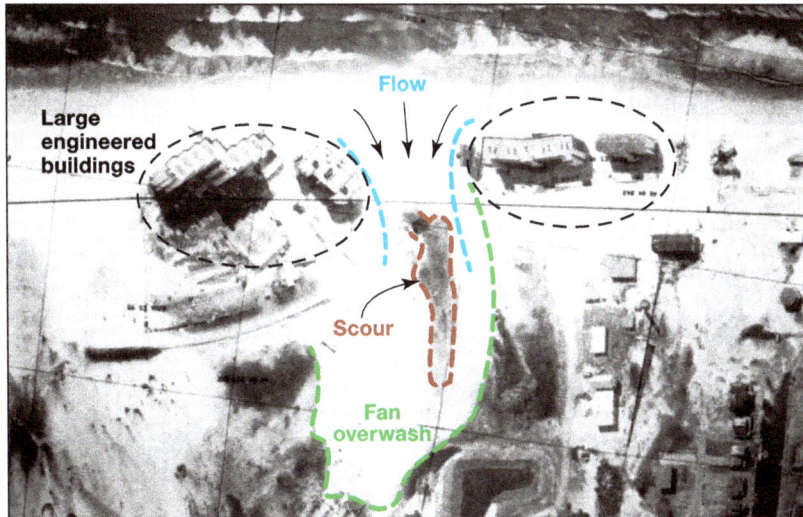

Figure 3-21.
Pile-supported house in the area of channeled flow shown in Figure 3-20. The building foundation and elevation successfully prevented high-velocity flow, erosion, and scour from destroying this building

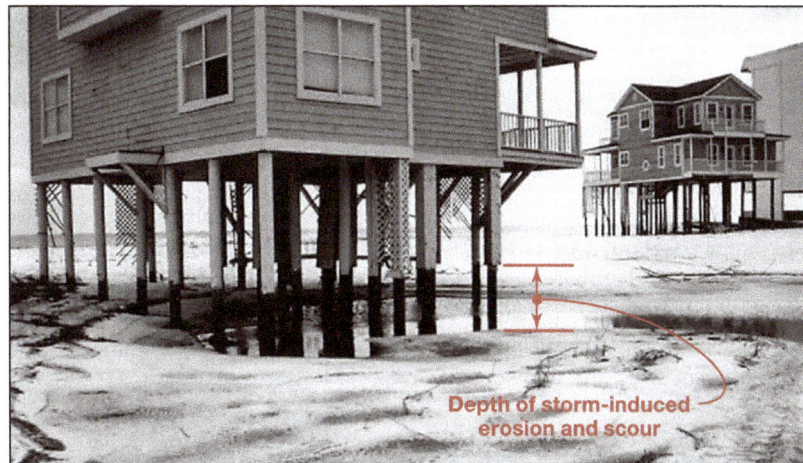

Figure 3-22.
This house, located in an area of channeled flow near that shown in Figure 3-20, was undermined, washed into the bay behind the barrier island, and became a threat to navigation

3.4.3   Waves

Waves can affect coastal buildings in a number of ways, including breaking waves, wave runup, wave reflection and deflection, and wave uplift. The most severe damage is caused by ***breaking waves*** (see Figure 3-23). The force created by waves breaking against a vertical surface is often 10 or more times higher than the force created by high winds during a storm event.

Figure 3-23.
Storm waves breaking against a seawall in front of a coastal residence at Stinson Beach, CA
SOURCE: LESLEY EWING, USED WITH PERMISSION

Wave runup occurs as waves break and run up beaches, sloping surfaces, and vertical surfaces. Wave runup (see Figure 3-24) can drive large volumes of water against or around coastal buildings, inducing fluid impact forces (albeit smaller than breaking wave forces), current drag forces, and localized erosion and scour (see Figure 3-25). Wave runup against a vertical wall generally extends to a higher elevation than runup on a sloping surface and is capable of destroying overhanging decks and porches. ***Wave reflection*** or ***deflection*** from adjacent structures or objects can produce forces similar to those caused by wave runup.

Figure 3-24.
Wave runup beneath elevated buildings at Scituate, MA, during the December 1992 nor'easter storm
SOURCE: JIM O'CONNELL, USED WITH PERMISSION

Wave runup

Shoaling waves beneath elevated buildings can lead to *wave uplift* forces. The most common example of wave uplift damage occurs at fishing piers, where pier decks are commonly lost close to shore, when shoaling storm waves lift the pier deck from the pilings and beams. The same type of damage can sometimes be observed at the lowest floor of insufficiently elevated, but well-founded, residential buildings and underneath slabs-on-grade below elevated buildings (see Figure 3-26).

Figure 3-25.
The sand underneath this Pensacola Beach, FL, building was eroded due to wave runup and storm surge (Hurricane Ivan, 2004)

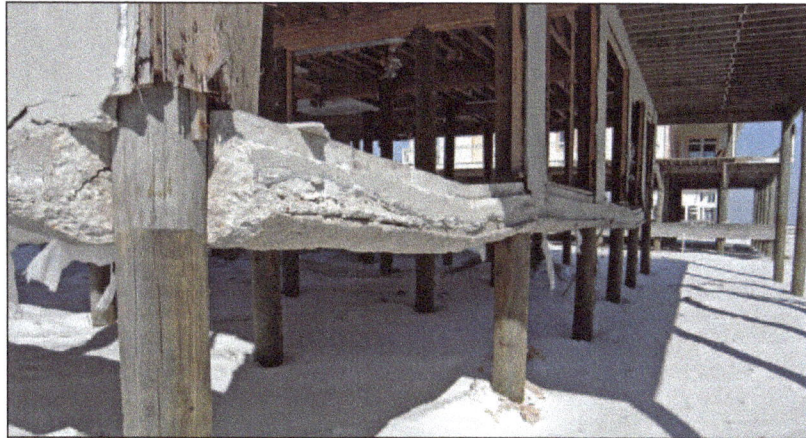

Figure 3-26.
Concrete slab-on-grade flipped up by wave action came to rest against two foundation members, generating large unanticipated loads on the building foundation (Topsail Island, NC, Hurricane Fran, 1996)

3.4.4   Flood-Borne Debris

Flood-borne debris produced by coastal flood events and storms typically includes decks, steps, ramps, breakaway wall panels, portions of or entire houses (see Figure 3-27), heating oil and propane tanks, vehicles, boats, decks and pilings from piers (see Figure 3-28), fences, destroyed erosion control structures, and a variety of smaller objects. Flood-borne debris is often capable of destroying unreinforced masonry walls, light wood-frame construction, and small-diameter posts and piles (and the components of structures they support). Figure 3-29 shows debris generated by destroyed buildings at Pass Christian, MS, that accumulated approximately 1,000 feet inland from the highway. The debris from buildings closest to the Gulf of Mexico undoubtedly accentuated damage to buildings in the area and contributed to their destruction. Debris trapped by cross bracing, closely spaced pilings, grade beams, or other components or obstructions below the BFE is also capable of transferring flood and wave loads to the foundation of an elevated structure. Parts of the country are exposed to more massive debris, such as the drift logs shown in Figure 3-30.

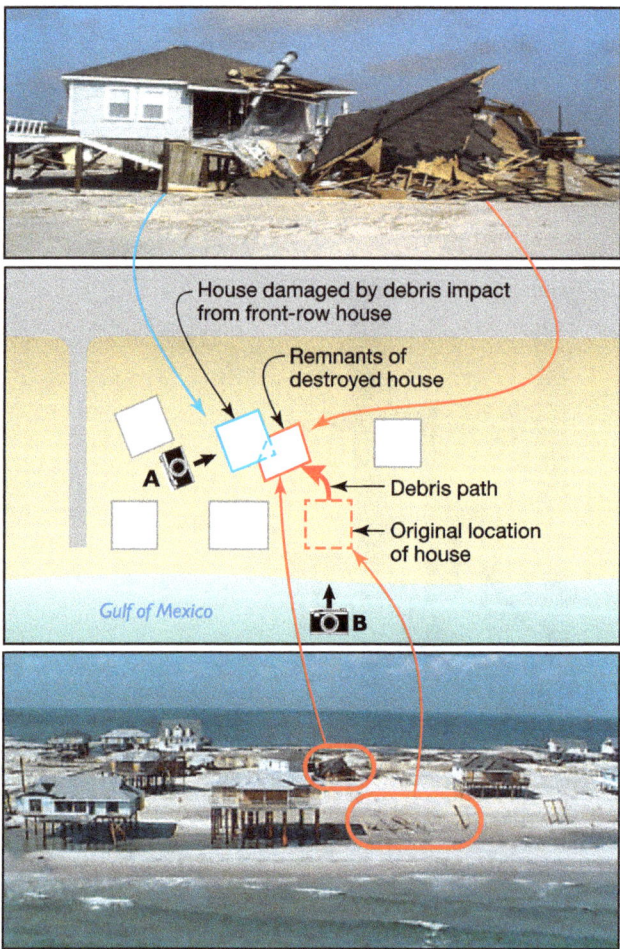

Figure 3-27.
A pile-supported house at Dauphin Island, AL, was toppled and washed into another house, which suffered extensive damage (Hurricane Georges, 1998)

Figure 3-28.
Pier pilings were carried over 2 miles by storm surge and waves before they came to rest against this elevated house in Pensacola Beach, FL (Hurricane Opal, 1995)

Figure 3-29.
Debris generated by destroyed buildings at Pass Christian, MS (Hurricane Katrina, 2005)

Reduced structural damage due to inundation and small waves (large waves attenuated by debris piles)

Severe structural damage due to waves and debris

Only floor slabs remain

Highway 90

Gulf of Mexico

Figure 3-30.
Drift logs driven into coastal houses at Sandy Point, WA, during a March 1975 storm

SOURCE: KNOWLES AND TERICH 1977, *SHORE AND BEACH*, USED WITH PERMISSION

3.5   Erosion

Erosion refers to the wearing or washing away of coastal lands. Although the concept of erosion is simple, erosion is one of the most complex hazards to understand and predict at a given site. Therefore, designers should develop an understanding of erosion fundamentals, but rely on coastal erosion experts (at Federal, State, and local agencies; universities; and private firms) for specific guidance regarding erosion potential at a site.

The term "erosion" is commonly used to refer to the horizontal recession of the shore (i.e., *shore erosion*), but can apply to other types of erosion. For example, *seabed* or *lakebed erosion* (also called *downcutting*) occurs when fine-grained sediments in the nearshore zone are eroded and carried into deep water. These sediments are lost permanently, resulting in a lowering of the seabed or lakebed. This process has several important consequences: increased local water depths, increased wave heights reaching the shoreline, increased shore erosion, and undermining of erosion control structures. Downcutting has been documented along some ocean-facing shorelines, but also along much of the Great Lakes shoreline

NOTE

This section reviews basic concepts related to coastal erosion, but cannot provide a comprehensive treatment of the many aspects of erosion that should be considered in planning, siting, and designing coastal residential buildings.

NOTE

Erosion is one of the most complex hazards faced by designers. However, given erosion data provided by experts, assessing erosion effects on building design can be reduced to three basic steps:

1. Define the most landward shoreline location expected during the life of the building.

2. Define the lowest expected ground elevation during the life of the building.

3. Define the highest expected BFE during the life of the building.

(which is largely composed of fine-grained glacial deposits). Designers should refer to Keillor (1998) for more information on this topic.

Erosion is capable of threatening coastal residential buildings in a number of ways:

- Destroying dunes or other natural protective features (see Figure 3-31)

- Destroying erosion control devices (see Figure 3-32)

- Lowering ground elevations, undermining shallow foundations, and reducing penetration depth of pile foundations (see Figure 3-33)

- Transporting beach and dune sediments landward, where they can bury roads and buildings and marshes (see Figure 3-34)

- Breaching low-lying coastal barrier islands exposing structures on the mainland to increased flood and wave effects (see Figures 3-35 and 3-36)

- Eroding coastal bluffs that provide support to buildings outside the floodplain itself (see Figure 3-37)

Sand that is moved during erosional events can create overwash and sediment burial issues. Further, the potential for landslides and ground failures must also be considered.

Figure 3-31.
Dune erosion in Ocean City, NJ, caused by the remnants of Hurricane Ida (2009) and a previous nor'easter

Figure 3-32.
Erosion and seawall damage in New Smyrna Beach, FL, following Hurricane Jeanne in 2007

Figure 3-33.
Erosion undermining a coastal residence in Oak Island, NC, caused by Hurricane Floyd in 1999

Figure 3-34.
Overwash on Topsail
Island, NC, after
Hurricane Bonnie in 1998
SOURCE: USGS

Figure 3-35. A January
1987 nor'easter cut a
breach across Nauset
Spit on Cape Cod,
MA; the breach grew
from an initial width of
approximately 20 feet
to over a mile within
2 years, exposing the
previously sheltered
shoreline of Chatham to
ocean waves and erosion
SOURCE: JIM O'CONNELL,
USED WITH PERMISSION

Figure 3-36.
Undermined house at
Chatham, MA, in 1988;
nine houses were lost as
a result of the formation
of the new tidal inlet
shown in Figure 3-35
SOURCE: JIM O'CONNELL,
USED WITH PERMISSION

Figure 3-37.
Bluff failure by a
combination of marine,
terrestrial, and seismic
processes led to
progressive undercutting
of blufftop apartments
at Capitola, CA, where
six of the units were
demolished after the
1989 Loma Prieta
earthquake
SOURCE: GRIGGS 1994,
*JOURNAL OF COASTAL
RESEARCH*, USED WITH
PERMISSION

3.5.1 Describing and Measuring Erosion

Erosion should be considered part of the larger process of shoreline change. When more sediment leaves a shoreline segment than moves into it, *erosion* results; when more sediment moves into a shoreline segment than leaves it, *accretion* results; and when the amounts of sediment moving into and leaving a shoreline segment balance, the shoreline is said to be *stable*.

Care must be exercised in classifying a particular shoreline as erosional, accretional, or stable. A shoreline classified as erosional may experience periods of stability or accretion. Likewise, a shoreline classified as stable or accretional may be subject to periods of erosion. Observed shoreline behavior depends on the time period of analysis and on prevailing and extreme coastal processes during that period.

For these reasons, shoreline changes are classified as short-term changes and long-term changes. Short-term changes occur over periods ranging from a few days to a few years and can be highly variable in direction and magnitude. Long-term changes occur over a period of decades, during which short-term changes tend to average out to the underlying erosion or accretion trend. Both short-term and long-term shoreline changes should be considered in siting and design of coastal residential construction.

Erosion is usually expressed as a rate, in terms of:

- Linear retreat (e.g., feet of shoreline recession per year)
- Volumetric loss (e.g., cubic yards of eroded sediment per foot of shoreline frontage per year)

The convention used in this Manual is to cite erosion rates as positive numbers, with corresponding shoreline change rates as negative numbers (e.g., an erosion rate of 2 feet per year is equivalent to a shoreline change rate of -2 feet per year). Likewise, accretion rates are listed as positive numbers, with corresponding shoreline change rates as positive numbers (e.g., an accretion rate of 2 feet per year is equivalent to a shoreline change rate of 2 feet per year).

Shoreline erosion rates are usually computed and cited as long-term, average annual rates. However, erosion rates are not uniform in time or space. Erosion rates can vary substantially from one location along the shoreline to another, even when the two locations are only a short distance apart.

A study by Zhang (1998) examined long-term erosion rates along the east coast of the United States. Results showed the dominant trend along the east coast of the United States is

NOTE

Most owners and designers worry only about erosion. However, sediment deposition and burial can also be a problem if dunes and windblown sand migrate inland.

NOTE

Short-term erosion rates can exceed long-term rates by a factor of 10 or more.

WARNING

Proper planning, siting, and design of coastal residential buildings require: (1) a basic understanding of shoreline erosion processes, (2) erosion rate information from the community, State, or other sources, (3) appreciation for the uncertainty associated with the prediction of future shoreline positions, and (4) knowledge that siting a building immediately landward of a regulatory coastal setback line does not guarantee the building will be safe from erosion. Owners and designers should also be aware that shore changes and modifications near to or updrift of a building site can affect the site.

one of erosion (72 percent of the stations examined experienced long-term erosion), with shoreline change rates averaging -3.0 feet per year (i.e., 3.0 feet per year of erosion). However, variability along the shoreline is considerable, with a few locations experiencing more than 20 feet per year of erosion, and over one-fourth of the stations experiencing accretion. A study of the Pacific County, WA, coastline found erosion rates as high as 150 feet per year, and accretion rates as high as 18 feet per year (Kaminsky et al. 1999).

Erosion rates can also vary over time at a single location. For example, Figure 3-38 illustrates the shoreline history over a period of 160 years for the region approximately 1.5 miles south of Indian River Inlet, DE. Although the long-term, average annual shoreline change rate is approximately -2 feet per year, short-term shoreline change rates vary from -27 feet per year (erosion resulting from severe storms) to +6 feet per year (accretion associated with post-storm recovery of the shoreline). This conclusion—that erosion rates can vary widely over time—has also been demonstrated by other studies (e.g., Douglas, et al., 1998).

Designers should also be aware that some shorelines experience large seasonal fluctuations in beach width and elevation. These changes are a result of seasonal variations in wave conditions and water levels, and should not be taken as indicators of long-term shoreline changes. For this reason, shoreline change calculations at beaches subject to large seasonal fluctuations should be based on shoreline measurements taken at approximately the same time of year.

NOTE

Apparent erosion or accretion resulting from seasonal fluctuations of the shoreline is not an indication of true shoreline change.

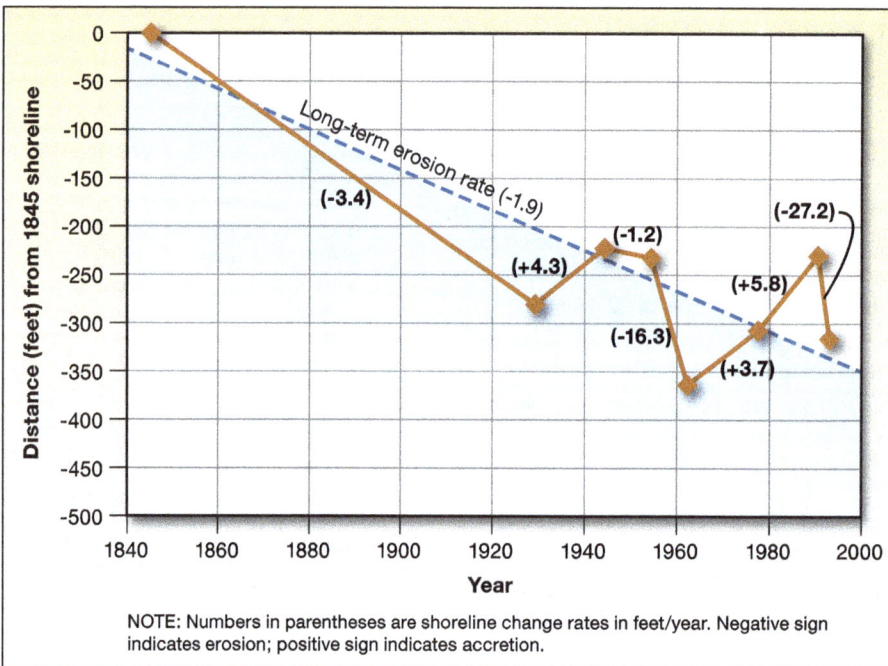

NOTE: Numbers in parentheses are shoreline change rates in feet/year. Negative sign indicates erosion; positive sign indicates accretion.

Figure 3-38.
Shoreline changes through time at a location approximately 1.5 miles south of Indian River Inlet, DE
DATA SOURCES: NOAA AND THE STATE OF DELAWARE

Erosion rates have been calculated by many States and communities to establish regulatory construction setback lines. These rates are typically calculated from measurements made with aerial photographs, historical charts, or beach profiles. However, a number of potential errors are associated with measurements and calculations using each of the data sources, particularly the older data. Some studies have estimated that errors in computed erosion rates can range up to 1 foot or more per year. Therefore, even if published erosion rates are less than 1 foot per year *this Manual recommends siting coastal residential structures based on the larger of the published erosion rate, or 1 foot per year,* unless there is compelling evidence to support a smaller erosion rate. Basing design on erosion rates of less than 1 foot per year can lead to significant underestimation of the future shoreline and inadequate setback to protect the building from long-term erosion.

3.5.2 Causes of Erosion

Erosion can be caused by a variety of natural or manmade actions, including:

- Storms and coastal flood events, usually rapid and dramatic (also called storm-induced erosion)

- Natural changes associated with tidal inlets, river outlets, and entrances to bays (e.g., interruption of littoral transport by jetties and channels, migration or fluctuation of channels and shoals, formation of new inlets)

- Construction of manmade structures and human activities (e.g., certain shore protection structures; damming of rivers; dredging or mining sand from beaches and dunes; and alteration of vegetation, surface drainage, or groundwater at coastal bluffs)

- Long-term erosion that occurs over a period of decades, due to the cumulative effects of many factors, including changes in water level, sediment supply, and those factors mentioned above

- Local scour around structural elements, including piles and foundation elements

CROSS REFERENCE

Chapters 12 and 13 provide information about designing and constructing sound pile and column foundations.

Erosion can affect all coastal landforms except highly resistant geologic formations. Low-lying beaches and dunes are vulnerable to erosion, as are most coastal bluffs, banks, and cliffs. Improperly sited buildings—even those situated atop coastal bluffs and outside the floodplain—and buildings with inadequate foundation support are especially vulnerable to the effects of erosion.

3.5.2.1 Erosion During Storms

Erosion during storms can be dramatic and damaging. Although storm-induced erosion is usually short-lived (usually occurring over a few hours in the case of hurricanes and typhoons, or over a few tidal cycles or days in the case of nor'easters and other coastal storms), the resulting erosion can be equivalent to decades of long-term erosion. During severe storms or coastal flood events, large dunes may be eroded 25 to 75 feet or more (see Figure 3-31) and small dunes may be completely destroyed.

Erosion during storms sometimes occurs despite the presence of erosion control devices such as seawalls, revetments, and toe protection. Storm waves frequently overtop, damage, or destroy poorly designed, constructed, or maintained erosion control devices. Lands and buildings situated behind an erosion control device are not necessarily safe from coastal flood forces and storm-induced erosion.

Narrow sand spits, barrier islands and low-lying coastal lands can be breached by tidal channels and inlets—often originating from the buildup of water on the back side (see Figure 3-39)—or washed away entirely (see Figure 3-40). Storm-induced erosion damage to unconsolidated cliffs and bluffs typically takes the form of large-scale collapse, slumping, and landslides, with concurrent recession of the top of the bluff.

CROSS REFERENCE

FIRMs incorporate the effects of dune and bluff erosion during storms (see Section 3.6.7).

Figure 3-39.
Breach through barrier island at Pine Beach, AL, before Hurricane Ivan (2001) and after (2004)
SOURCE: USGS

Figure 3-40.
Cape San Blas, Gulf County, FL, in November 1984, before and after storm-induced erosion

Storm-induced erosion can take place along open-coast shorelines (Atlantic, Pacific, Gulf of Mexico, and Great Lakes shorelines) and along shorelines of smaller enclosed or semi-enclosed bodies of water. If a body of water is subject to increases in water levels and generation of damaging wave action during storms, storm-induced erosion can occur.

3.5.2.2   Erosion Near Tidal Inlets, Harbor, Bay, and River Entrances

Many miles of coastal shoreline are situated on or adjacent to connections between two bodies of water. These connections can take the form of tidal inlets (short, narrow hydraulic connections between oceans and inland waters), harbor entrances, bay entrances, and river entrances. The size, location, and adjacent shoreline stability of these connections are usually governed by six factors:

WARNING

The location of a tidal inlet, harbor entrance, bay entrance, or river entrance can be stabilized by jetties or other structures, but the shorelines in the vicinity can still fluctuate in response to storms, waves, and other factors.

- Tidal and freshwater flows through the connection
- Wave climate
- Sediment supply
- Local geology
- Jetties or stabilization structures
- Channel dredging

Temporary or permanent changes in any of these governing factors can cause the connections to migrate, change size, or change configuration, and can cause sediment transport patterns in the vicinity of the inlet to change, thereby altering flood hazards in nearby areas.

Construction of jetties or similar structures at a tidal inlet or a bay, harbor, or river entrance often results in accretion on one side and erosion on the other, with a substantial shoreline offset. This offset results from the jetties trapping the *littoral drift* (wave-driven sediment moving along the shoreline) and preventing it from moving to the downdrift side. Figure 3-41 shows such a situation at Ocean City Inlet, MD, where formation

Figure 3-41.
Ocean City Inlet, MD, was opened by a hurricane in 1933 and stabilized by jetties in 1934–35 that have resulted in extreme shoreline offset and downdrift erosion (1992 photograph)

of the inlet in 1933 by a hurricane and construction of inlet jetties in 1934–1935 led to approximately 800 feet of accretion against the north jetty at Ocean City and approximately 1,700 feet of erosion on the south side of the inlet along Assateague Island as of 1977 (Dean and Perlin 1977). Between 1976 and 1980, shoreline change rates on Assateague Island averaged from 49 feet per year and -33 feet per year (USACE 2009b). In 2004, USACE began the "Long-Term Sand Management" project to restore Assateague Island.

Erosion and accretion patterns at stabilized inlets and entrances sometimes differ from the classic pattern occurring at the Ocean City Inlet. In some instances, accretion occurs immediately adjacent to both jetties, with erosion beyond. In some instances, erosion and accretion patterns near a stabilized inlet change over time. Figure 3-42 shows buildings at Ocean Shores, WA, that were threatened by shore erosion shortly after their construction, despite the fact that the buildings were located near an inlet jetty on a beach that was historically viewed as accretional.

Development in the vicinity of a tidal inlet or bay, harbor, or river entrance is often affected by lateral migration of the channel and associated changes in sand bars (which may focus waves and erosion on particular shoreline areas). Often, these changes are cyclic in nature and can be identified and forecast through a review of historical aerial photographs and bathymetric data. Those considering a building site near a tidal inlet or a bay, harbor, or river entrance should investigate the history of the connection, associated shoreline fluctuations, migration trends, and impacts of any stabilization structures. Failure to do so could result in increased building vulnerability or building loss to future shoreline changes.

> **NOTE**
>
> Cursory characterizations of shoreline behavior in the vicinity of a stabilized inlet, harbor, or bay entrance should be rejected in favor of a more detailed evaluation of shoreline changes and trends.

> **WARNING**
>
> Many State and local siting regulations allow residential development in areas where erosion is likely to occur. Designers should not assume that a building sited in compliance with minimum State and local requirements is safe from future erosion. See Chapter 4.

Figure 3-42. Buildings threatened by erosion at Ocean Shores, WA, in 1998. The rock revetments were built in response to shore erosion along an area adjacent to a jetty and thought to be accretional

Shoreline changes in the vicinity of one of the more notable regulatory takings cases illustrate this point. The upper image in Figure 3-43 is a 1989 photograph of one of the two vacant lots owned by David Lucas, which became the subject of the *Lucas vs. South Carolina Coastal Council* case when Lucas challenged the State's prohibition of construction on the lots. By December 1997, the case had been decided in favor of Lucas, the State of South Carolina had purchased the lots from Lucas, the State had resold the lots, and a home had been constructed on one of the lots (Jones et al. 1998). The lower image in Figure 3-43 shows a December 1997 photograph of the same area, with erosion undermining the home built on the former Lucas lot (left side of photograph) and an adjacent house (also present in 1989 in upper image).

Figure 3-43.
July 1989 photograph of vacant lot owned by Lucas, Isle of Palms, SC (top) and photograph taken in December 1997 of lot with new home (bottom)

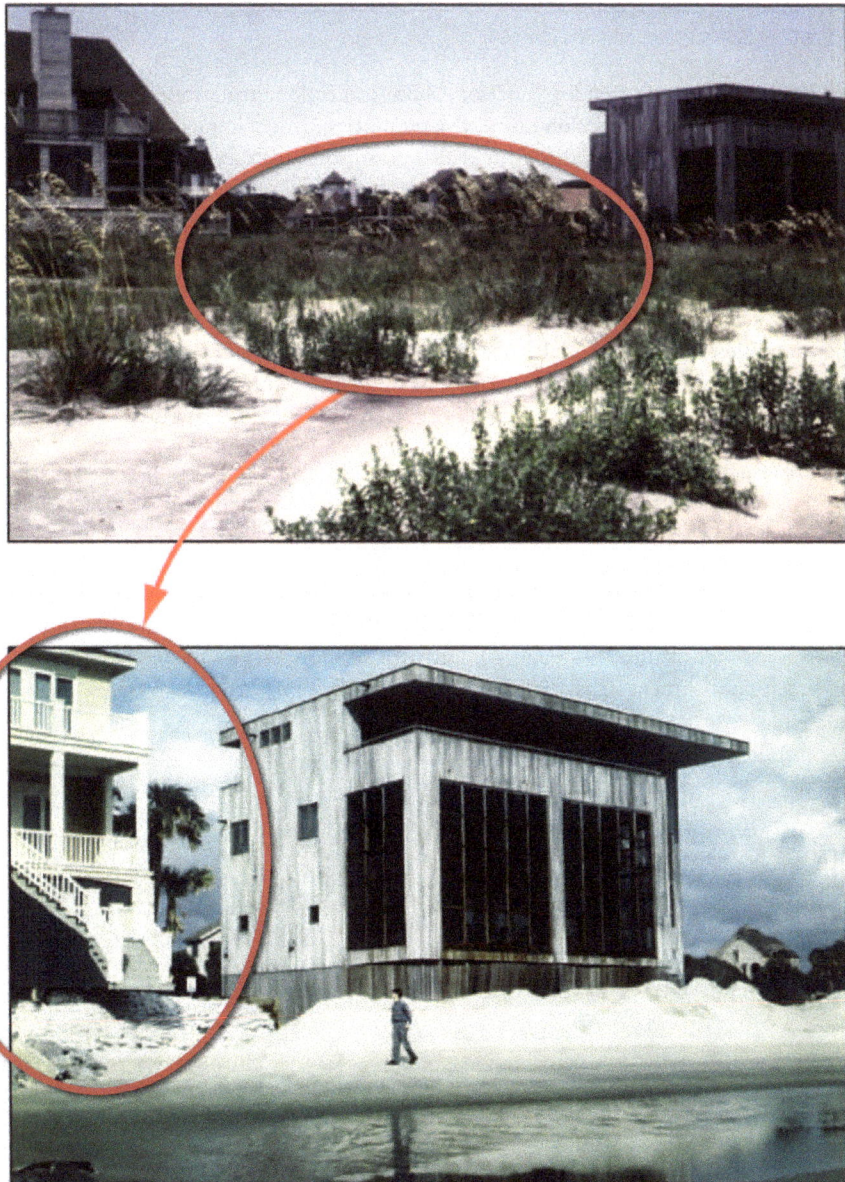

3.5.2.3 Erosion Due to Manmade Structures and Human Activities

Human actions along the shoreline can both reduce and increase flood hazards. In some instances, structures built or actions taken to facilitate navigation cause erosion elsewhere. In other cases, structures built or actions taken to halt erosion and reduce flood hazards at one site increase erosion and flood hazards at nearby sites. For this reason, evaluation of a potential coastal building site requires consideration of natural and human-caused shoreline changes.

NOTE

More information on beach nourishment is provided at http://www.csc.noaa.gov/ beachnourishment.

Effects of Shore Protection Structures

In performing their intended function, shore protection structures can lead to or increase erosion on nearby properties. This statement should not be taken as an indictment of all erosion control structures, because many provide protection against erosion and flood hazards. Rather, this Manual simply recognizes the potential for adverse impacts of these structures on nearby properties and offers some siting guidance for residential buildings relative to erosion control structures (see Section 4.6), where permitted by States and communities. These potential impacts vary from site to site and structure to structure and can sometimes be mitigated by ***beach nourishment***—the placement of additional sediment on the beach—in the vicinity of the erosion control structure.

CROSS REFERENCE

Adverse impacts of erosion control structures can sometimes be mitigated through beach nourishment. See Section 4.7.

Groins (such as those shown in Figure 2-12, in Chapter 2) are short, shore-perpendicular structures designed to trap available littoral sediments. They can cause erosion to downdrift beaches if the groin compartments are not filled with sand and maintained in a full condition.

Likewise, ***offshore breakwaters*** (see Figure 3-44) can trap available littoral sediments and reduce the sediment supply to nearby beaches. This adverse effect should be mitigated by combining breakwater construction with beach nourishment—design guidance for offshore breakwater projects typically calls for the inclusion of beach nourishment (Chasten et al. 1993).

Figure 3-44.
Example of littoral sediments being trapped behind offshore breakwaters on Lake Erie, Presque Isle, PA
SOURCE: USACE

Seawalls, bulkheads, and *revetments* are shore-parallel structures built, usually along the shoreline or at the base of a bluff, to act as retaining walls and to provide some degree of protection against high water levels, waves, and erosion. The degree of protection they afford depends on their design, construction, and maintenance. They do not prevent erosion of the beach, and in fact, can exacerbate ongoing erosion of the beach. The structures can impound upland sediments that would otherwise erode and nourish the beach, lead to *passive erosion* (eventual loss of the beach as a structure prevents landward migration of the beach profile), and lead to active erosion (localized scour waterward of the structure and on unprotected property at the ends of the structure).

Post-storm inspections show that the vast majority of privately financed seawalls, revetments, and erosion control devices fail during 1-percent-annual-chance, or lesser, events (i.e., are heavily damaged or destroyed, or withstand the storm, but fail to prevent flood damage to lands and buildings they are intended to protect—see Figures 3-32 and 3-45). Reliance on these devices to protect inland sites and residential buildings is not a good substitute for proper siting and foundation design. Guidance on evaluating the ability of existing seawalls and similar structures to withstand a 1-percent-annual-chance coastal flood event can be found in Walton et al. (1989).

Finally, some communities distinguish between erosion control structures constructed to protect existing development and those constructed to create a buildable area on an otherwise unbuildable site. Designers should investigate any local or State regulations and requirements pertaining to erosion control structures before selecting a site and undertaking building design.

Effects of Alteration of Vegetation, Drainage, or Groundwater

Alteration of vegetation, drainage, or groundwater can sometimes make a site more vulnerable to coastal storm or flood events. For example, removal of vegetation (grasses, ground covers, trees, mangroves) at a site can render the soil more prone to erosion by wind, rain, and flood forces. Alteration of natural drainage patterns

> ⚠️ **WARNING**
>
> NFIP regulations require that communities protect mangrove stands in Zone V from any human-caused alteration that would increase potential flood damage.

Figure 3-45.
Failure of seawall in Bay County, FL, led to undermining and collapse of the building behind the wall (Hurricane Opal, 1995)

and groundwater flow can lead to increased erosion potential, especially on steep slopes and coastal bluffs. Irrigation and septic systems often contribute to bluff instability problems by elevating groundwater levels and decreasing soil strength.

3.5.2.4 Long-Term Erosion

Observed long-term erosion at a site represents the net effect of a combination of factors. The factors that contribute to long-term erosion can include:

- Sea level rise or subsidence of uplands

- Lake level rise or lakebed erosion along the Great Lakes (Figure 3-46)

- Reduced sediment supply to the coast

- Construction of jetties, other structures, or dredged channels that impede littoral transport of sediments along the shoreline

- Increased incidence or intensity of storms

- Alteration of upland vegetation, drainage, or groundwater flows (especially in coastal bluff areas)

WARNING

Coastal FIRMs (even recently published coastal FIRMs) do not incorporate the effects of long-term erosion. Users are cautioned that mapped Zone V and Zone A areas subject to long-term erosion underestimate the extent and magnitude of actual flood hazards that a coastal building may experience over its lifetime.

Regardless of the cause, long-term shore erosion can increase the vulnerability of coastal construction in a number of ways, depending on local shoreline characteristics, construction setbacks, and structure design. Figure 3-47 shows an entire block of buildings that are dangerously close to the shoreline and vulnerable to storm damage due to the effects of long-term erosion.

Figure 3-46. Long-term erosion of the bluff along the Lake Michigan shoreline in Ozaukee County, WI, increases the threat to residential buildings outside the floodplain (1996 photograph)

Figure 3-47.
Long-term erosion at
South Bethany Beach,
DE, has lowered ground
elevations beneath
buildings and left them
more vulnerable to
storm damage

SOURCE: CHRIS JONES
1992, USED WITH
PERMISSION

In essence, *long-term erosion acts to shift flood hazard zones landward.* For example, a site mapped accurately as Zone A may become exposed to Zone V conditions; a site accurately mapped as outside the 100-year floodplain may become exposed to Zone A or Zone V conditions.

Despite the fact that FIRMs do not incorporate long-term erosion, other sources of long-term erosion data are available for much of the country's shorelines. These data usually take the form of historical shoreline maps or erosion rates published by individual States or specific reports (from Federal or State agencies, universities, or consultants) pertaining to counties or other small shoreline reaches.

Designers should be aware that more than one source of long-term erosion rate data may be available for a given site and that the different sources may report different erosion rates. Differences in rates may be a result of different study periods, different data sources (e.g., aerial photographs, maps, ground surveys), or different study methods. When multiple sources and long-term erosion rates exist for a given site, designers should use the highest long-term erosion rate in their siting decisions, unless they conduct a detailed review of the erosion rate studies and conclude that a lower erosion rate is more appropriate for forecasting future shoreline positions.

3.5.2.5   Localized Scour

Localized scour can occur when water flows at high velocities past an object embedded in or resting on erodible soil (localized scour can also be caused or exacerbated by waves interacting with the object). The scour is not caused by the flood or storm event, per se, but by the distortion of the flow field by the object; localized scour occurs only around the object itself and is in addition to storm- or flood-induced erosion that occurs in the general area.

Flow moving past a fixed object must accelerate, often forming eddies or vortices and scouring loose sediment from the immediate vicinity of the object. Localized scour around individual piles and similar objects (see Figure 3-48) is generally limited to small, cone-shaped depressions (less than 2 feet deep and several feet in diameter). Localized scour is capable of undermining slabs and grade-supported structures. However, in severe cases, the depth and lateral extent of localized scour can be much greater, and will jeopardize foundations and may lead to structural failure. Figure 3-49 shows severe local scour that occurred around residential foundations on Bolivar Peninsula, TX, after Hurricane Ike in 2008. This type of scour was widespread during Hurricane Ike. Although some structures were able to withstand the scour and associated flood forces, others were not.

> **CROSS REFERENCE**
>
> Refer to Section 8.5 for additional discussion on scour.

Designers should consider potential effects of localized scour when calculating foundation size, depth, or embedment requirements.

Figure 3-48. Determination of localized scour from changes in sand color, texture, and bedding (Hurricane Fran, 1996)

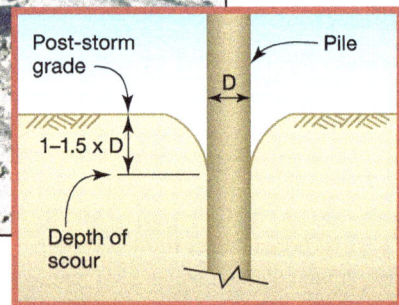

Figure 3-49.
Residential foundation
that suffered severe
scour on Bolivar
Peninsula, TX (Hurricane
Ike, 2008)

3.5.3   Overwash and Sediment Burial

Sediment eroded during a coastal storm event must travel to
one of the following locations: offshore to deeper water, along
the shoreline, or inland. Overwash occurs when low-lying
coastal lands are overtopped and eroded by storm surge and
waves, such that the eroded sediments are carried landward by
floodwaters, burying uplands, roads, and at-grade structures
(see Figure 3-50). Depths of overwash deposits can reach 3 to
5 feet, or more, near the shoreline, but gradually decrease with
increasing distance from the shoreline. Overwash deposits can extend several hundred feet inland following
a severe storm (see Figure 3-34), especially in the vicinity of shore-perpendicular roads. Post-storm aerial
photographs and/or videos can be used to identify likely future overwash locations.

> **NOTE**
> Most owners and designers worry
> only about erosion. However,
> sediment deposition and burial
> can also be a problem.

The physical processes required to create significant overwash deposits (i.e., waves capable of suspending
sediments in the water column and flow velocities generally in excess of 3 feet per second) are also capable of
damaging buildings. Thus, existing coastal buildings located in Zone A (particularly the seaward portions
of Zone A) and built on slab or crawlspace foundations should be considered vulnerable to damage from
overwash, high-velocity flows, and waves.

3.5.4   Landslides and Ground Failures

Landslides occur when slopes become unstable and loose material slides or flows under the influence of gravity.
Often, landslides are triggered by other events such as erosion at the toe of a steep slope, earthquakes, floods,
or heavy rains, but can be worsened by human actions such as destruction of vegetation or uncontrolled
pedestrian access on steep slopes (see Figure 3-51). An extreme example is Hurricane Mitch in 1998, where
heavy rainfall led to flash flooding, numerous landslides, and an estimated 10,000 deaths in Nicaragua.

Figure 3-50.
Overwash from Hurricane Opal (1995) at Pensacola Beach, FL, moved sand landward from the beach and buried the road, adjacent lots, and some at-grade buildings to a depth of 3 to 4 feet

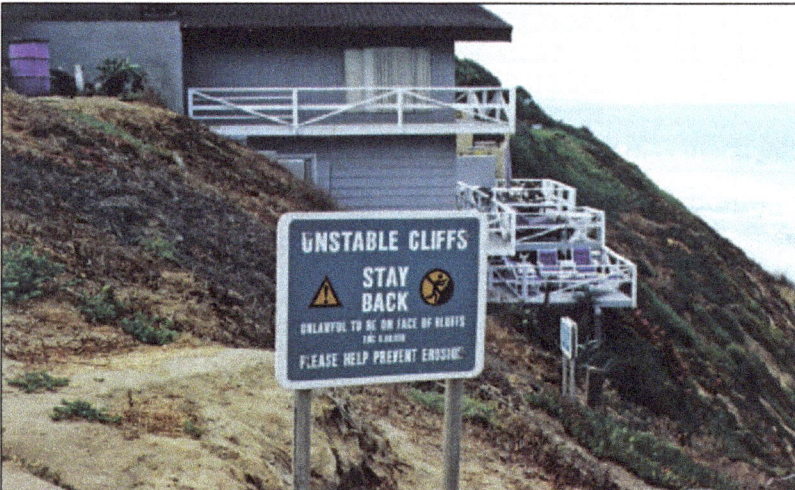

Figure 3-51.
Unstable coastal bluff at Beacon's Beach, San Diego, CA
SOURCE: LESLEY EWING, USED WITH PERMISSION

Designers should seek and use landslide information and data from State geological survey agencies and USGS (http://landslides.usgs.gov/). Designers should also be aware that coastal bluff failures can be induced by seismic activity. Griggs and Scholar (1997) detail bluff failures and damage to residential buildings resulting from several earthquakes, including the March 1964 Alaska earthquake and the October 1989 Loma Prieta earthquake (see Figure 3-37). Coastal bluff failures were documented as far away as 50 miles from the Loma Prieta epicenter and 125 miles from the Alaska earthquake epicenter. In both instances, houses and infrastructure were damaged and destroyed as a result of these failures.

3.6   NFIP Flood Hazard Zones

Understanding the methods and assumptions underlying Flood Insurance Study (FIS) reports and FIRMs is useful to the designer, especially in the case where the effective FIRM is more than a few years old, and where an updated flood hazard determination is desired.

FEMA determines flood hazards at a given site based on the following factors:

- Anticipated flood conditions (stillwater elevation, wave setup, wave runup and overtopping, and wave propagation) during the base flood event (based on the flood level that has a 1-percent chance of being equaled or exceeded in any given year)

- Potential for storm-induced erosion of the primary dune during the base flood event

- Physical characteristics of the floodplain, such as vegetation and existing development

- Topographic and bathymetric information

> **NOTE**
>
> A detailed discussion of the methodology for computing stillwater elevations, wave heights, and wave runup is beyond the scope of this Manual. Refer to *Guidelines and Specifications for Flood Hazard Mapping Partners* (FEMA 2003) for more information.

- Computer models are used to calculate flood hazards and water surface elevations. FEMA uses the results of these analyses to map BFEs and flood hazard zones.

3.6.1 Base Flood Elevations

To determine BFEs for areas affected by coastal flooding, FEMA computes 100-year *stillwater elevations* and *wave setup*, and then determines the maximum 100-year *wave heights* and, in some areas, the maximum 100-year *wave runup*, associated with those stillwater elevations. Wave heights are the heights, above the wave trough, of the crests of wind-driven waves. Wave runup is the rush of wave water up a slope or structure. Stillwater elevations are the elevations of the water surface resulting solely from storm surge (i.e., the rise in the surface of the ocean due to the action of wind and the drop in atmospheric pressure associated with hurricanes and other storms).

> **NOTE**
>
> Note that rounding of coastal BFEs means that it is possible for the wave crest or wave runup elevation to be up to 0.5 foot above the lowest floor elevation. This is another reason to incorporate freeboard into design.

The stillwater elevation plus wave setup equals the *mean water elevation*, which serves as the surface across which waves propagate. Several factors can contribute to the 100-year mean water elevation in a coastal area. The most important factors include offshore bathymetry, astronomical tide, wind setup (rise in water surface as strong winds blow water toward the shore), pressure setup (rise in water surface due to low atmospheric pressure), wave setup (rise in water surface inside the surf zone due to the presence of breaking waves), and, in the case of the Great Lakes, seiches and variations in lake levels.

The BFEs shown for coastal flood hazard areas on FIRMs are established not at the stillwater elevation, but at the elevation of either the wave crest or the wave runup (rounded to the nearest foot), whichever is greater. Whether the wave crest elevation or the wave runup elevation is greater depends primarily on upland topography. In general, wave crest elevations are greater where the upland topography is gentle, such as along most of the Gulf, southern Atlantic, and middle-Atlantic coasts, while wave runup elevations are greater where the topography is steeper, such as along portions of the Great Lakes, northern Atlantic, and Pacific coasts.

3.6.2   Flood Insurance Zones

The insurance zone designations shown on FIRMs indicate the magnitude and severity of flood hazards. The zone designations that apply to coastal flood hazard areas are listed below, in decreasing order of magnitude and severity.

Zones VE, V1–V30, and V. These zones, collectively referred to as Zone V, identify the Coastal High Hazard Area, which is the portion of the SFHA that extends from offshore to the inland limit of a ***primary frontal dune*** along an open coast and any other portion of the SFHA that is subject to high-velocity wave action from storms or seismic sources. The boundary of Zone V is generally based on wave heights (3 feet or greater) or wave runup depths (3 feet or greater). Zone V can also be mapped based on the ***wave overtopping*** rate (when waves run up and over a dune or barrier).

Zones AE, A1–A30, AO, and A. These zones, collectively referred to as Zone A or AE, identify portions of the SFHA that are not within the Coastal High Hazard Area. Zones AE, A1–A30, AO, and A are used to designate both coastal and non-coastal SFHAs. Regulatory requirements of the NFIP for buildings located in Zone A are the same for both coastal and riverine flooding hazards.

Limit of Moderate Wave Action (LiMWA). Zone AE in coastal areas is divided by the LiMWA. The LiMWA represents the landward limit of the 1.5-foot wave. The area between the LiMWA and the Zone V limit is known as the ***Coastal A Zone*** for building code and standard purposes and as the ***Moderate Wave Action (MoWA)*** area by FEMA flood mappers. This area is subject to wave heights between 1.5 and 3 feet during the base flood. The area between the LiMWA and the landward limit of Zone A due to coastal flooding is known as the ***Minimal Wave Action*** (***MiWA***) area, and is subject to wave heights less than 1.5 feet during the base flood.

Zones X, B, and C. These zones identify areas outside the SFHA. Zone B and shaded Zone X-500 identify areas subject to inundation by the flood that has a 0.2-percent chance of being equaled or exceeded during any given year, often referred to as the 500-year flood. Zone C and unshaded Zone X identify areas outside the 500-year floodplain. Areas protected by accredited levee systems are mapped as shaded Zone X.

TERMINOLOGY

SPECIAL FLOOD HAZARD AREA (SFHA) defines an area with a 1-percent chance, or greater, of flooding in any given year. This is commonly referred to as the extent of the 100-year floodplain.

COASTAL SFHA is the portion of the SFHA where the source of flooding is coastal surge or inundation. It includes Zone VE and Coastal A Zone.

ZONE VE is that portion of the coastal SFHA where base flood wave heights are 3 feet or greater, or where other damaging base flood wave effects have been identified, or where the primary frontal dune has been identified.

COASTAL A ZONE (MoWA AREA) is that portion of the coastal SFHA referenced by building codes and standards, where base flood wave heights are between 1.5 and 3 feet, and where wave characteristics are deemed sufficient to damage many NFIP-compliant structures on shallow or solid wall foundations.

MiWA AREA is that portion of the Coastal SFHA where base flood wave heights are less than 1.5 feet.

LiMWA is the boundary between the MoWA and the MiWA.

RIVERINE SFHA is that portion of the SFHA mapped as Zone AE and where the source of flooding is riverine, not coastal.

ZONE AE is the portion of the SFHA not mapped as Zone VE. It includes the MoWA, the MiWA, and the Riverine SFHA.

3.6.3   FIRMs, DFIRMs, and FISs

Figure 3-52 shows a typical paper FIRM that a designer might encounter for some coastal areas. Three flood hazard zones are shown on this FIRM: Zone V, Zone A, and Zone X. Figure 3-53 shows an example of a transect perpendicular to the shoreline.

Since the early 2000s, FEMA has been preparing Digital FIRMs (DFIRMs) to replace the paper maps. Figure 3-54 shows a typical DFIRM that a designer is likely to encounter in many coastal areas. The DFIRM uses a photographic base and shows either the results of a recent FIS or the results of a digitized paper FIRM (possibly with a datum conversion from National Geodetic Vertical Datum [NGVD] to North American Vertical Datum [NAVD]). The flood hazard zones and BFEs on a DFIRM are delineated in a manner consistent with those on a paper FIRM, although they may reflect updated flood hazard calculation procedures.

CROSS REFERENCE

See Section 3.3 for a brief discussion of coastal flood hazards and FIRMs.

NOTE

Additional information about FIRMs is available in FEMA's 2006 booklet *How to Use a Flood Map to Protect Your Property*, FEMA 258 (FEMA 2006b).

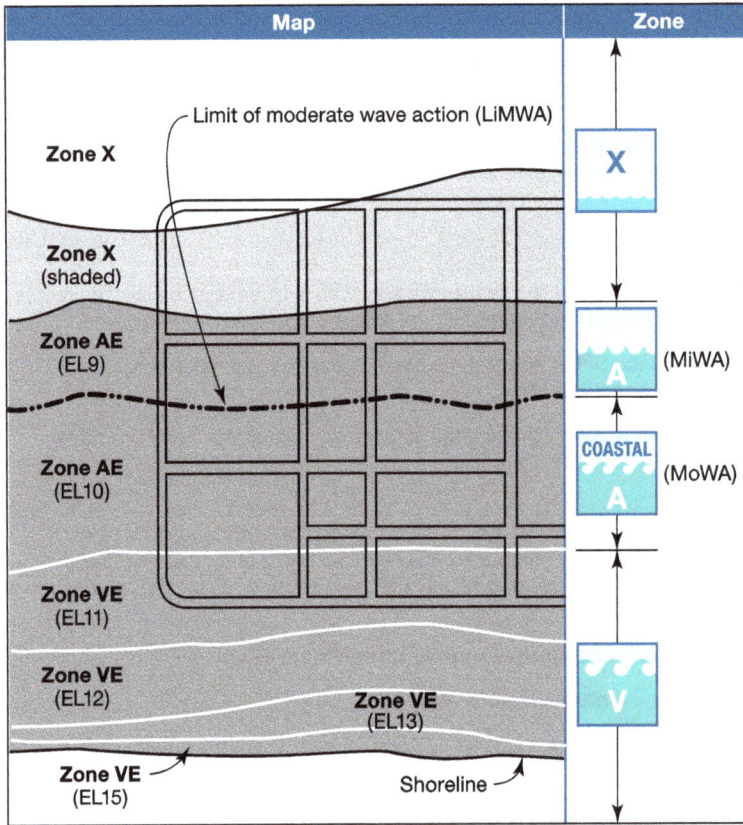

Figure 3-52.
Portion of a paper FIRM showing coastal flood insurance rate zones. The icons on the right indicate the associated flood hazard zones for design and construction purposes. The LiMWA is not shown on older FIRMs, but is shown on newer FIRMs and DFIRMs

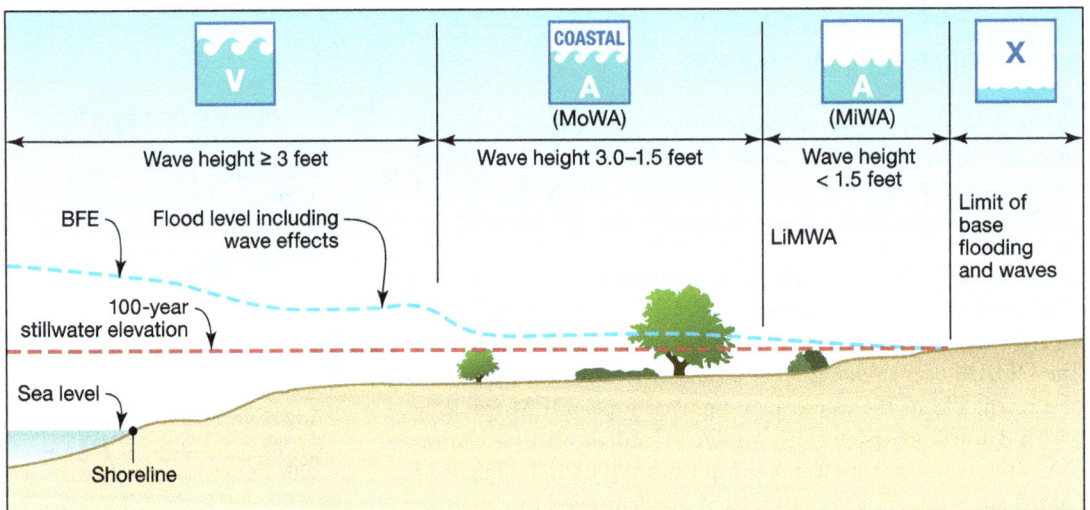

Figure 3-53.
Typical shoreline-perpendicular transect showing stillwater and wave crest elevations and associated flood zones

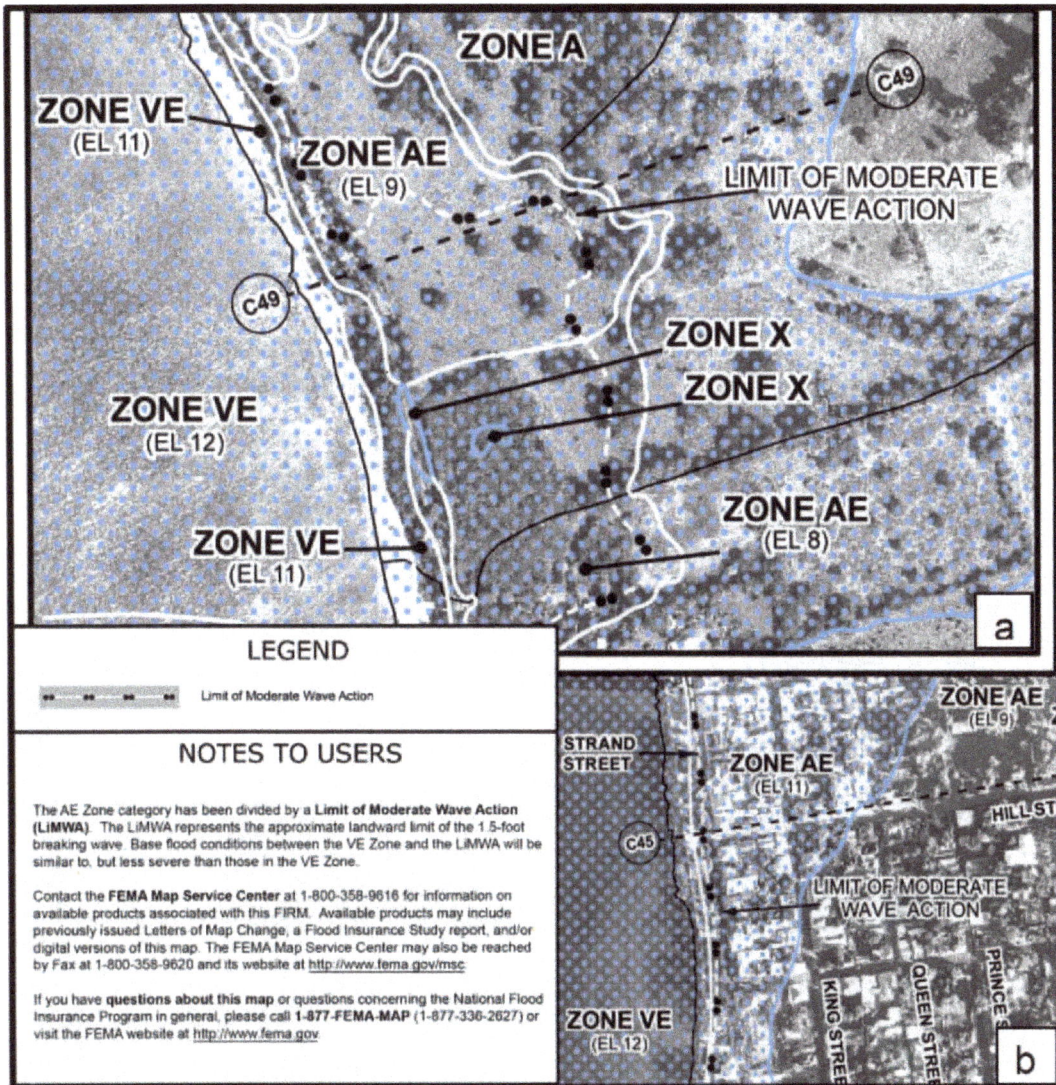

Figure 3-54.
Example DFIRM for a coastal area that shows the LiMWA
SOURCE: FEMA 2008c

A coastal FIS is completed with FEMA-specified techniques and procedures (see FEMA 2007) to determine mean water levels (stillwater elevation plus wave setup) and wave elevations along transects drawn perpendicular to the shoreline (see Figure 3-53). The determination of the 100-year mean water elevation (and elevations associated with other return intervals) is usually accomplished through the statistical analysis of historical tide and water level data, and/or by the use of numerical storm surge and wave models. Wave heights and elevations on land are computed from mean water level and topographic data with established procedures and models that account for wave dissipation by obstructions (e.g., sand dunes, buildings, vegetation) and wave regeneration across overland fetches.

Building codes and standards—and FEMA building science publications—refer to the Coastal A Zone and have specific requirements or recommendations for design and construction in this zone. Post-disaster damage inspections consistently show the need for such a distinction. Figure 3-53 shows how the Coastal A Zone can be inferred from FIS transects and maps.

NOTE

Detailed FEMA coastal mapping guidance is contained in Appendix D of *Guidelines and Specifications for Flood Hazard Mapping Partners* (FEMA 2003). Designers need not be familiar with all of these guidelines, but they may be useful on occasion. Appendix D is divided into several documents, one for the Atlantic and Gulf of Mexico coasts, one for the Pacific coast, and one for the Great Lakes coast. These documents have been and continue to be updated and revised, so designers should refer to the FEMA mapping Web site for the latest versions: http://www.fema.gov/plan/prevent/fhm/dl_vzn.shtm#3. Guidance on mapping the LiMWA is contained in *Procedure Memorandum No. 50* at http://www.fema.gov/library/viewRecord.do?id=3481.

3.6.4   Wave Heights and Wave Crest Elevations

FEMA's primary means of establishing BFEs and distinguishing between Zone V, Zone A, and Zone X is ***wave height***. Wave height is simply the vertical distance between the crest and trough of a wave propagating over the water surface. BFEs in coastal areas are usually set at the elevation of the crest of the wave as it propagates inland.

The maximum wave crest elevation (used to establish the BFE) is determined by the maximum wave height, which depends

TERMINOLOGY: WAVE HEIGHT

Wave height is the vertical distance between the wave crest and wave trough (see Figure 3-55). Wave crest elevation is the elevation of the crest of a wave, referenced to the NGVD, NAVD, or other datum.

largely on the 100-year stillwater depth (d_{100}). This depth is the difference between the 100-year stillwater elevation (E_{100}) (including wave setup) and the ground elevation (noted as *GS* in Figure 3-55). Note that ground elevation in this use is ***not*** the existing ground elevation, but is the ground elevation that will result from the erosion expected to occur during the base flood (or in some cases, it may be appropriate to take it as the eroded ground elevation expected over the life of a building).

In shallow waters the maximum height of a breaking wave (H_b) is usually taken to be 78 percent of the stillwater depth d_s, and determined by the equation $H_b = 0.78d_s$. However, designers should be aware that where steep slopes exist immediately seaward of a building, wave heights can exceed $0.78d_{sw}$ (and a reasonable alternative is to set $H_b = 1.00d_s$ in such instances).

The wave form in shallow water is distorted so that the crest and trough are not equidistant from the stillwater level; for NFIP flood mapping purposes, the wave crest lies at 70 percent of the wave height above the stillwater elevation (the wave trough lies a distance equal to 30 percent of the wave height, below the stillwater elevation). Thus, the maximum elevation of a breaking wave crest above the stillwater elevation is equal to

CROSS REFERENCE

See Equation 8.1 and Example 8.1 for calculations pertaining to stillwater depth (d_s).

$0.55d_s$. In the case of the 1-percent-annual-chance (base) flood, $H_b = 0.78d_{100}$ and the maximum height of a breaking wave above the 100-year stillwater elevation = $0.55d_{100}$ (see Figure 3-55). Note that for wind-driven waves, water depth is only one of three parameters that determine the actual wave height at a particular site (wind speed and fetch length are the other two). In some instances, actual wave heights may be below the depth-limited maximum height.

Figure 3-55. BFE determination for coastal flood hazard areas where wave crest elevations exceed wave runup elevations (Zones A and V)

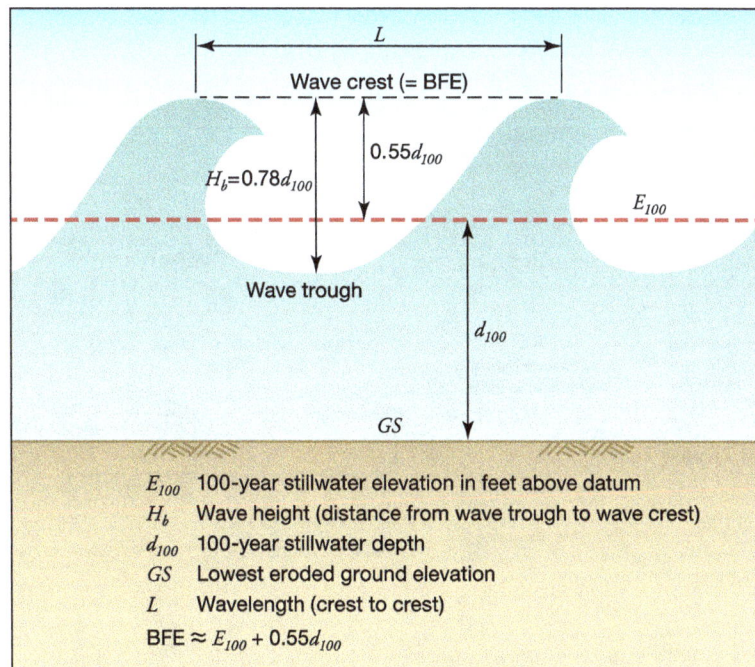

For a coastal flood hazard area where the ground slopes up gently from the shoreline, and there are few obstructions such as houses and vegetation, the BFE shown on the FIRM is approximately equal to the ground elevation plus the 100-year stillwater depth (d_{100}) plus $0.55d_{100}$. For example, where the ground elevation is 4 feet NAVD and d_{100} is 6 feet, the BFE is equal to 4 feet plus 6 feet plus 3.3 feet, or 13.3 feet NAVD, rounded to 13 feet NAVD.

3.6.5   Wave Runup

On steeply sloped shorelines, the rush of water up the surface of the natural beach (including dunes and bluffs) or the surface of a manmade structure (such as a revetment or vertical wall) can result in flood elevations higher than those of the crests of wind-driven waves. For a coastal flood hazard area where this situation occurs, the BFE shown on the FIRM is equal to the highest elevation reached by the water (see Figure 3-56).

3.6.6   Primary Frontal Dune

The NFIP has other parameters used to establish Zone V delineations besides wave heights and wave runup depths. In some cases, the landward limit of the primary frontal dune will determine the landward limit of Zone V. This Zone V designation is based on dune morphology, as opposed to base flood conditions. Consult the *Guidelines and Specifications for Flood Hazard Mapping Partners* (FEMA 2003) for details regarding the NFIP primary frontal dune delineation. Note that some States and communities may have different dune definitions, but these will not be used by the NFIP to map Zone V.

NOTE

FEMA maps Zone V based on *wave heights* where the wave height (vertical distance between wave crest and wave trough) is greater than or equal to 3 feet.

NOTE

FEMA maps Zone V based on *wave runup* where the vertical distance between the runup elevation and the ground (the runup "depth") is greater than or equal to 3 feet.

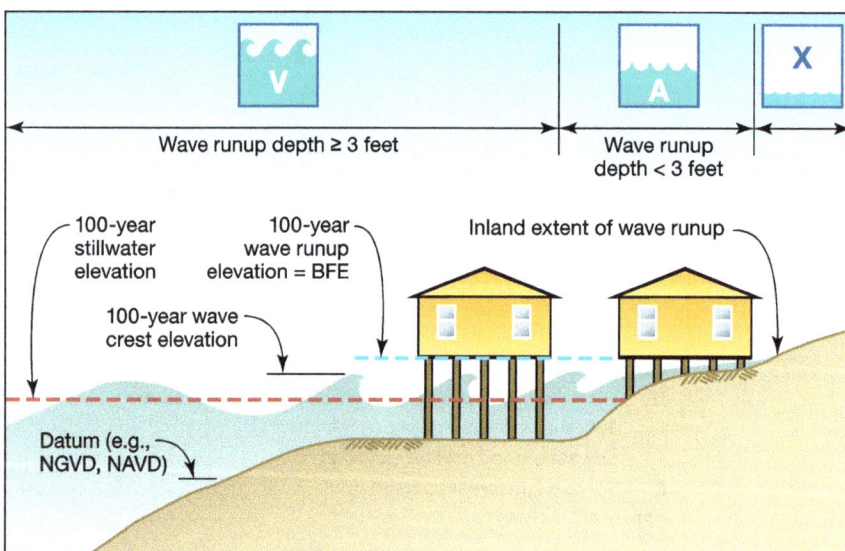

Figure 3-56. Where wave runup elevations exceed wave crest elevations, the BFE is equal to the runup elevation

> **TERMINOLOGY**
>
> **WAVE RUNUP** is the rush of water up a slope or structure.
>
> **WAVE RUNUP DEPTH** at any point is equal to the maximum wave runup elevation minus the lowest eroded ground elevation at that point.
>
> **WAVE RUNUP ELEVATION** is the elevation reached by wave runup, referenced to NGVD or other datum.
>
> **WAVE SETUP** is an increase in the stillwater surface elevation near the shoreline, due to the presence of breaking waves. Wave setup typically adds 1.5 to 2.5 feet to the 100-year stillwater flood elevation.
>
> **MEAN WATER ELEVATION** is the sum of the stillwater elevation and wave setup.

3.6.7 Erosion Considerations and Flood Hazard Mapping

Proper design requires two types of erosion to be considered: dune and bluff erosion during the base flood event, and long-term erosion. Newer FIRMs account for the former, but no FIRMs account for the latter.

Dune/Bluff Erosion. Current FIS procedures account for the potential loss of protective dunes and bluffs during the 100-year flood. However, this factor was not considered in coastal FIRMs prepared prior to May 1988, which delineated Zone V without any consideration for storm-induced erosion. Zone V boundaries were drawn at the crest of the dune solely on the basis of the elevation of the ground and without regard for the erosion that would occur during a storm.

Long-Term Erosion. Designers, property owners, and floodplain managers should be careful not to assume that flood hazard zones shown on FIRMs accurately reflect current flood hazards, especially if there has been a significant natural hazard event since the FIRM was published. For example, flood hazard restudies completed after Hurricane Opal (1995, Florida Panhandle) and Fran (1996, Topsail Island, NC) have produced FIRMs that are dramatically different from the FIRMs in effect prior to the hurricanes.

Figure 3-57 provides an example of the effects of both dune erosion and long-term erosion changes. The figure compares pre- and post-storm FIRMs for Surf City, NC. The map changes are attributable to two factors: (1) pre-storm FIRMs did not show the effects of erosion that occurred after the FIRMs were published and did not meet technical standards currently in place, and (2) Hurricane Fran caused significant changes to the topography of the barrier island. Not all coastal FIRMs would be expected to undergo such drastic revisions after a flood restudy; however, many FIRMs may be in need of updating, and designers should be aware that FIRMs may not accurately reflect present flood hazards at a site.

3.6.8 Dune Erosion Procedures

Current Zone V mapping procedures (FEMA 2003) require that a dune have a minimum *frontal dune reservoir* (dune cross-section above 100-year stillwater level and seaward of dune peak) of 540 square feet in order to be considered substantial enough to withstand erosion during a base flood event. According to FEMA procedures, a frontal dune reservoir less than 540 square feet will result in dune removal (dune disintegration), while a frontal dune reservoir greater than or equal to 540 square feet generally will result in dune retreat (see Figure 3-58).

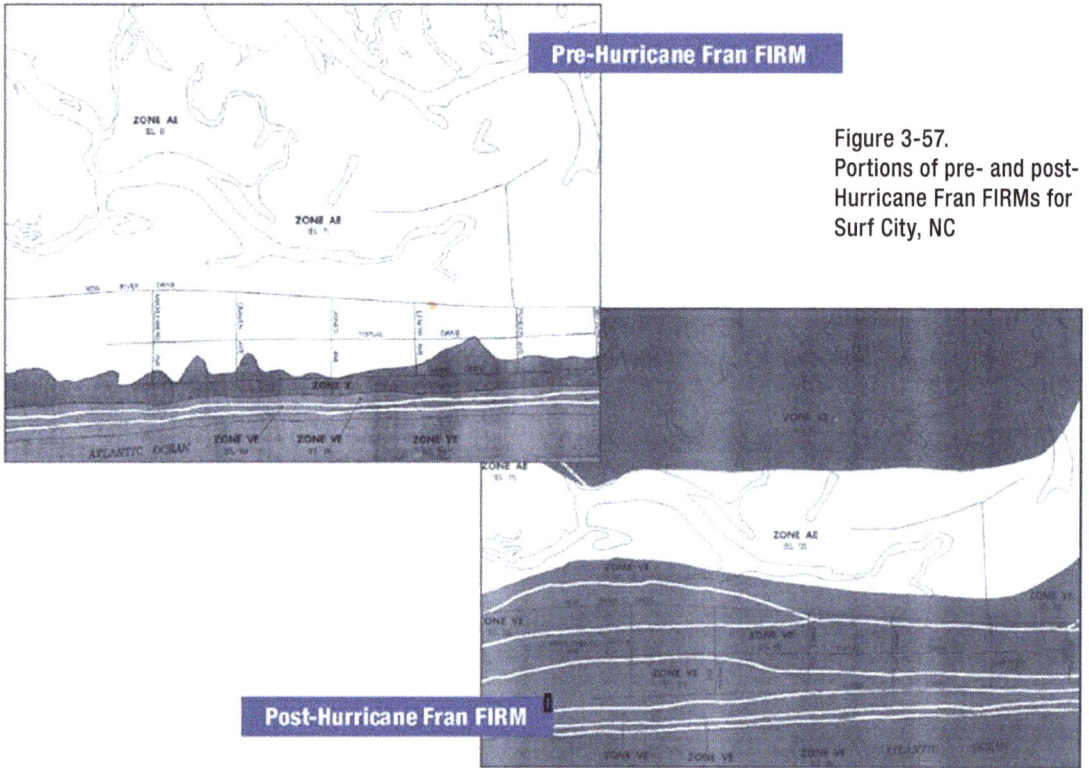

Figure 3-57.
Portions of pre- and post-
Hurricane Fran FIRMs for
Surf City, NC

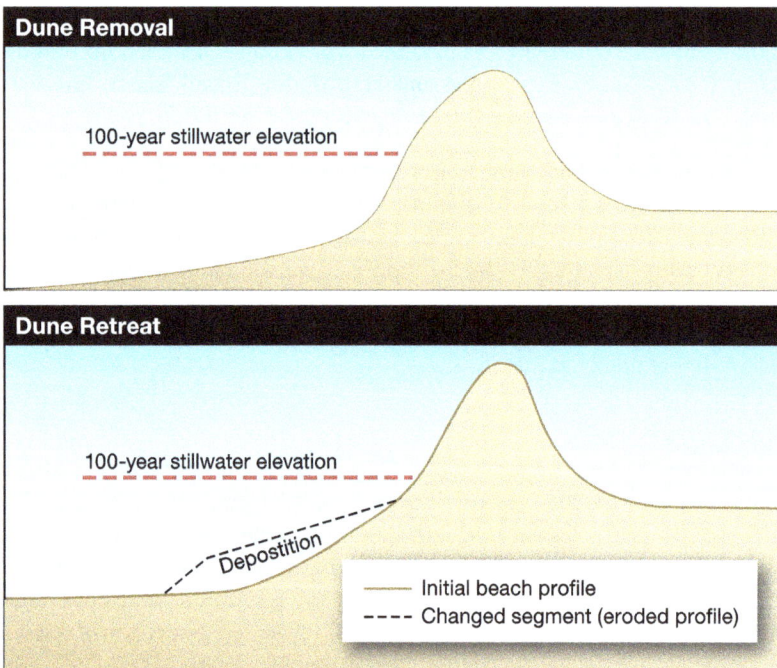

Figure 3-58.
Current FEMA treatment
of dune removal and dune
retreat
SOURCE: FEMA 2003

The current procedure for calculating the post-storm profile in the case of dune removal is relatively simple: a straight line is drawn from the pre-storm dune toe landward at an upward slope of 1 on 50 (vertical to horizontal) until it intersects the pre-storm topography landward of the dune. Any sediment above the line is assumed to be eroded.

This Manual recommends that the size of the frontal dune reservoir used by designers to prevent dune removal during a 100-year storm be increased to 1,100 square feet. This recommendation is made for three reasons: (1) The 540 square feet rule used by FEMA reflects dune size at the time of mapping and does not account for future conditions, when beaches and dunes may be compromised by long-term erosion; (2) The 540 square feet rule does not account for the cumulative effects of multiple storms that may occur within short periods of time, such as in 1996, when Hurricanes Bertha and Fran struck the North Carolina coast within 2 months of each other (see Figure 4-6 in Chapter 4); and (3) even absent long-term erosion and multiple storms, use of the median frontal dune reservoir underestimates dune erosion 50 percent of the time.

Dune erosion calculations at a site should also take **dune condition** into account. A dune that is not covered by well-established vegetation (i.e., vegetation that has been in place for two or more growing seasons) is more vulnerable to wind and flood damage than one with well-established vegetation. A dune crossed by a road or pedestrian path offers a weak point that storm waves and flooding exploit; to reduce potential weak points, elevated dune walkways are recommended. Post-storm damage inspections frequently show that dunes are breached at these weak points and structures landward of them are more vulnerable to erosion and flood damage.

3.6.9 Levees and Levee Protection

The floodplain area landward of a levee system for which the levee system provides a certain level of risk reduction is known as the **levee-impacted area.** Some levees include interior drainage systems that provide for conveyance of outflow of streams and runoff. Levee-impacted areas protected by accredited levees meeting NFIP requirements are mapped as Zone X (shaded) and the interior drainage areas are designated

> **CROSS REFERENCE**
>
> Section 2.6.2 provides additional detail on the risks of siting a building in a levee-impacted area.

as Zone A. For levees not meeting NFIP requirements, both sides of the levee are mapped as Zone A. Levees on older FIRMs may not have been evaluated against NFIP criteria, and may not offer the designed level of protection due to deterioration, changed hydrology or channel characteristics, or partial levee failure.

3.7 Flood Hazard Assessments for Design Purposes

Designers may sometimes be faced with a FIRM and FIS that are several years old, or older. As such, designers should determine whether the FIRM still accurately represents flood hazards associated with the site under present day base flood conditions. If not, the designer may need to pursue updating the information in order to more accurately understand the hazard conditions at the site.

> **WARNING**
>
> Some sites lie outside flood hazard areas shown on FIRMs, but may be subject to current or future flood and erosion hazards. These sites, like those within mapped flood hazard areas, should be evaluated carefully.

3.7.1   Determine If Updated or More Detailed Flood Hazard Assessment is Needed

Two initial questions drive the decision to update or complete a more detailed flood hazard assessment:

1. Does the FIRM accurately depict present flood hazards at the site of interest?

2. Will expected shore erosion render the flood hazard zones shown on the FIRM obsolete during the projected life of the building or development at the site?

The first question can be answered with a brief review of the FIRM, the accompanying FIS report, and site conditions. The answer to the second question depends upon whether or not the site is experiencing long-term shore erosion. If the shoreline at the site is stable and is not experiencing long-term erosion, then the FIRM does not require revision for erosion considerations. However, because FIRMs are currently produced without regard to long-term erosion, if a shoreline fluctuates or experiences long-term erosion, the FIRM will cease to provide the best available data at some point in the future (if it has not already) and a revised flood hazard assessment will be necessary.

Updated and revised flood hazard assessments are discussed with siting and design purposes in mind, not in the context of official changes to FIRMs that have been adopted by local communities. The official FEMA map change process is a separate issue that is not addressed by this Manual. Moreover, some siting and design recommendations contained in this Manual exceed minimum NFIP requirements, and are not tied to a community's adopted FIRM and its associated requirements.

3.7.1.1   Does the FIRM Accurately Depict Present Flood Hazards?

In order to determine whether a FIRM represents current flood hazards, and whether an updated or more detailed flood hazard assessment is needed, the following steps should be carried out:

> **NOTE**
>
> The date of the effective (i.e., newest) FIRM for a community can be found on FEMA's Web site under the heading "Community Status Book," at http://www.fema.gov/fema/csb.shtm.

- Obtain copies of the latest FIRM and FIS report for the site of interest. If the effective date precedes the critical milestones listed in Section 3.8, an updated flood hazard assessment may be needed.

- Review the legend on the FIRM to determine the history of the panel (and revisions to it), and review the study methods described in the FIS. If the revisions and study methods are not consistent with current study methods (FEMA 2007), an updated flood hazard assessment may be needed.

- If the FIS calculated dune erosion using the 540 square feet criterion (refer to Section 3.5.8) and placed the Zone V boundary on top of the dune, check the dune cross-section to see if it has a frontal dune reservoir of at least 1,100 square feet above the 100-year stillwater elevation. If not, consider shifting the Zone V boundary to the landward limit of the dune and revising other flood hazard zones, as needed.

- Review the description in the FIS report of the storm, water level, and flood source data used to generate the 100-year stillwater elevation and BFEs. If significant storms or flood events have affected the area since the FIS report and FIRM were completed, the source data may need to be revised and an updated flood hazard assessment may be needed.

- Determine whether there have been significant physical changes to the site since the FIS and FIRM were completed (e.g., erosion of dunes, bluffs, or other features; opening of a tidal inlet; modifications to drainage, groundwater, or vegetation on coastal bluffs; construction or removal of shore protection structures; filling or excavation of the site). If there have been significant changes in the physical configuration and condition since the FIS and FIRM were completed, an updated and more detailed flood hazard assessment may be needed.

- Determine whether adjacent properties have been significantly altered since the FIS and FIRM were completed (e.g., development, construction, excavation, etc.) that could affect, concentrate, or redirect flood hazards on the site of interest. If so, an updated and more detailed flood hazard assessment may be needed.

> **NOTE**
>
> Where a new FIRM exists (i.e., based on the most recent FEMA study procedures and topographic data), long-term erosion considerations can be approximated by shifting all flood hazard zones landward a distance equal to the long-term annual erosion rate multiplied by the life of the building or development (use 50 years as the minimum life). The shift in the flood hazard zones results from a landward shift of the profile.

If, after following the steps above, it is determined that an updated flood hazard assessment may be needed, see Section 3.7.2 for more information on updating and revising flood hazard assessments.

3.7.1.2   Will Long-Term Erosion Render a FIRM Obsolete?

Designers should determine whether a FIRM is likely to become obsolete as a result of long-term erosion considerations, and whether a revised flood hazard assessment is needed. First, check with local or State CZM agencies for any information on long-term erosion rates or construction setback lines. If such rates have been calculated, or if construction setback lines have been established from historical shoreline changes, long-term erosion considerations may necessitate a revised flood hazard assessment.

In cases where no long-term erosion rates have been published, and where no construction setback lines have been established based on historical shoreline movements, designers should determine whether the current shoreline has remained in the same approximate location as that shown on the FIRM (e.g., has there been any significant shore erosion, accretion, or fluctuation?). If there has been significant change in the shoreline location or orientation since the FIS and FIRM were completed, a revised flood hazard assessment may be needed.

3.7.1.3   Will Sea Level Rise Render a FIRM Obsolete?

Sea level rise has two principal effects: (1) it increases storm tide elevations and allows for larger wave heights to reach a coastal site, and (2) it leads to shoreline erosion. For these reasons, designers should investigate potential sea level rise and determine whether projected sea level changes will increase flood hazards at a site. Relying on the FIRM to project future site and base flood conditions may not be adequate in many cases. The NOAA site http://tidesandcurrents.noaa.gov/sltrends/sltrends.html provides historical information that a designer can extrapolate into the future. Designers may also wish to consider whether accelerated rates of rise will occur in the future.

A USACE Engineering Circular (USACE 2009a) provides guidance on sources of sea level change data and projections, and discusses how the data and projections can be used for planning purposes. The guidance is useful for planning and designing coastal residential buildings.

3.7.2   Updating or Revising Flood Hazard Assessments

Updating or revising an existing flood hazard assessment—for siting and design purposes—can be fairly simple or highly complex, depending upon the situation. A simple change may involve shifting a Zone A or Zone X boundary, based upon topographic data that is better than those used to generate the FIRM. A complex change may involve a detailed erosion assessment and significant changes to mapped flood hazard zones.

If an assessment requires recalculating local flood depths and wave conditions on a site, FEMA models (Erosion, Runup, and WHAFIS) can be used for the site (bearing in mind the recommended change to the required dune reservoir to prevent dune loss, described in 3.5.8).

If an assessment requires careful consideration of shore erosion, the checklist, flowchart, and diagram shown in Chapter 4 can be a guide, but a qualified coastal professional should be consulted. Much of the information and analyses described in the checklist and flowchart is likely to have already been developed and carried out previously by others, and should be available in reports about the area; designers are advised to check with the community. Cases for which information is unavailable and basic analyses have not been completed are rare.

The final result of the assessment should be a determination of the greatest flood hazards resulting from a 1-percent-annual-chance coastal flood event that the site will be exposed to over the anticipated life of a building or development. The determination should account for short- and long-term erosion, bluff stability, sea level rise, and storm-induced erosion; in other words, both chronic and catastrophic flood and erosion hazards, along with future water level conditions, should be considered.

3.8   Milestones of FEMA Coastal Flood Hazard Mapping Procedures and FIRMs

Designers are reminded that FEMA's flood hazard mapping procedures have evolved over the years (the coastal mapping site, http://www.fema.gov/plan/prevent/fhm/dl_vzn.shtm, provides links to current coastal mapping guidance and highlights many of these changes). Thus, a FIRM produced today might differ from an earlier FIRM, not only because of physical changes at the site, but also because of changes in FEMA hazard zone definitions, revised models, and updated storm data. Major milestones in the evolution of FEMA flood hazard mapping procedures, which can render early FIRMs obsolete, include:

- In approximately 1979, a FEMA storm surge model replaced NOAA tide frequency data as the source of storm tide stillwater elevations for the Atlantic and Gulf of Mexico coasts.

- In approximately 1988, coastal tide frequency data from the USACE New England District replaced earlier estimates of storm tide elevations for New England.

- In approximately 1988, return periods for Great Lakes water levels from the USACE Detroit District replaced earlier estimates of lake level return periods.

- There have been localized changes in flood elevations. For example, after Hurricane Opal (1995), a revised analysis of historical storm tide data in the Florida panhandle raised 100-year stillwater flood elevations and BFEs by several feet (Dewberry & Davis 1997).

- Prior to Hurricane Frederic in 1979, BFEs in coastal areas were set at the storm surge stillwater elevation, not at the wave crest elevation. Beginning in the early 1980s, FIRMs have been produced with Zone V, using the WHAFIS model and the 3-foot wave height as the landward limit of Zone V.

- Beginning in approximately 1980, tsunami hazard zones on the Pacific coast were mapped using procedures developed by the USACE. These procedures were revised in approximately 1995 for areas subject to both tsunami and hurricane effects.

- Before May 1988, flood hazard mapping for the Atlantic and Gulf of Mexico coasts was based solely on ground elevations and without regard for erosion that would occur during the base flood event; this practice resulted in Zone V boundaries being drawn near the crest of the primary frontal dune. Changes in mapping procedures in May 1988 accounted for storm-induced dune erosion and shifted many Zone V boundaries to the landward limit of the primary frontal dune.

- After approximately 1989, FIRMs were produced using a revised WHAFIS model, a runup model, and wave setup considerations to map flood hazard zones.

- Beginning in approximately 1989, a Great Lakes wave runup methodology (developed by the USACE Detroit District and modified by FEMA) was employed.

- Beginning in approximately 1989, a standardized procedure for evaluating coastal flood protection structures (Walton et al. 1989) was employed.

- Beginning in approximately 2005, FEMA began mapping the 2-percent exceedance wave runup elevation during the base flood instead of the mean runup elevation.

- In 2005, FEMA issued its *Final Draft Guidelines for Coastal Flood Hazard Analysis and Mapping for the Pacific Coast of the United States.*

- Beginning in 2005, FEMA began using advanced numerical storm surge (ADCIRC) and offshore wave (STWAVE and SWAN) models for Atlantic and Gulf of Mexico coastal flood insurance studies (conventional dune erosion procedures and WHAFIS are still used on land). Studies completed using these models should be considered the most accurate and reliable.

- In 2007, FEMA issued its *Atlantic Ocean and Gulf of Mexico Coastal Guidelines Update.*

In 2007, FEMA issued guidance for mapping the 500-year (0.2-percent-annual-chance) wave envelope in coastal studies.

In 2008, FEMA issued guidance for mapping coastal flood hazards in sheltered waters.

In December 2008, FEMA issued mapping guidance for the LiMWA (FEMA 2008c), which delineates the 1.5-foot wave height location, and thus, defines the landward limit of the Coastal A Zone.

In 2009, FEMA issued its *Great Lakes Coastal Guidelines Update* (FEMA 2009).

3.9   References

ASCE (American Society of Civil Engineers). 2010. *Minimum Design Loads for Buildings and Other Structures.* ASCE Standard ASCE 7-10.

Blake, E.S., E.N. Rappaport, J.D. Jarell, and C.W. Landsea. 2005. *The Deadliest, Costliest, and Most Intense United States Hurricanes from 1851 to 2004 (and Other Frequently Requested Hurricane Facts).* NOAA Technical Memorandum. NWS-TPC-4, 48 pp.

Caldwell, S. R.; R. D. Crissman. 1983. *Design for Ice Forces.* A State of the Practice Report. Technical Council on Cold Regions Engineering. American Society of Civil Engineers.

Camfield, F. E. 1980. *Tsunami Engineering.* Special Report No. 6. U.S. Army, Coastal Engineering Research Center.

Chasten, M. A.; J. D. Rosati; J. W. McCormick; R. E. Randall. 1993. *Engineering Design Guidance for Detached Breakwaters as Shoreline Stabilization Structures.* Technical Report CERC 93-13. U.S. Army Corps of Engineers, Coastal Engineering Research Center.

Chen, A. T.; C. B. Leidersdorf. 1988. *Arctic Coastal Processes and Slope Protection Design.* Monograph. Technical Council on Cold Regions Engineering. American Society of Civil Engineers.

Dean, R. G.; M. Perlin. 1977. "Coastal Engineering Study of Ocean City Inlet, Maryland." *Proceedings of the ASCE Specialty Conference, Coastal Sediments '77.* American Society of Civil Engineers. New York. pp. 520–542.

Dewberry & Davis, Inc. 1997. *Executive Summary of Draft Report, Coastal Flood Studies of the Florida Panhandle.*

Douglas, B. C.; M. Crowell; S. P. Leatherman. 1998. "Considerations for Shoreline Position Prediction." *Journal of Coastal Research.* Vol. 14, No. 3, pp. 1025–1033.

FEMA (Federal Emergency Management Agency). 1996. *Corrosion Protection for Metal Connectors in Coastal Areas for Structures Located in Special Flood Hazard Areas in accordance with the National Flood Insurance Program.* Technical Bulletin 8-96.

FEMA. 1997. *Multi-Hazard Identification and Risk Assessment, A Cornerstone of the National Mitigation Strategy.*

FEMA. 1998. *Wildfire Mitigation in the 1998 Florida Wildfires.* Wildfire Report FEMA-1223-DR-FL. FEMA, Region IV.

FEMA. 2003. *Guidelines and Specifications for Flood Hazard Mapping Partners.* April.

FEMA. 2005. *Final Draft Guidelines for Coastal Flood Hazard Analysis and Mapping for the Pacific Coast of the United States.* January.

FEMA. 2006a. *Homebuilders' Guide to Earthquake Resistant Design and Construction.* FEMA 232. June.

FEMA. 2006b. *How to Use a Flood Map to Protect Your Property.* FEMA 258.

FEMA. 2007. *Atlantic Ocean and Gulf of Mexico Coastal Guidelines Update. Final Draft.* February.

FEMA. 2008a. *Taking Shelter From the Storm: Building a Safe Room For Your Home or Small Business.* FEMA 320. August.

FEMA. 2008b. *Home Builder's Guide to Construction in Wildfire Zones.* FEMA P-737.

FEMA. 2008c. Procedure Memorandum No. 50 – *Policy and Procedures for Identifying and Mapping Areas Subject to Wave Heights Greater than 1.5 feet as an Informational Layer on Flood Insurance Rate Maps (FIRMs).*

FEMA. 2009. *Great Lakes Coastal Guidelines Update.* http://www.floodmaps.fema.gov/pdf/fhm/great_lakes_guidelines.pdf. March.

FEMA. 2010. Fact Sheet 1.7 "Coastal Building Materials." *Home Builder's Guide to Coastal Construction.* FEMA P-499. December.

Griggs, G. B. 1994. "California's Coastal Hazards." *Coastal Hazards Perception, Susceptibility and Mitigation,* Journal of Coastal Research Special Issue No. 12. C. Finkl, Jr., ed., pp. 1–15.

Griggs, G. B.; D. C. Scholar. 1997. Coastal Erosion Caused by Earthquake-Induced Slope Failure. *Shore and Beach.* Vol. 65, No. 4, pp. 2–7.

Harris-Galveston Subsidence District. 2010. "Subsidence 1906–2000." *Data Source National Geodetic Survey.* http://www.hgsubsidence.org/assets/pdfdocuments/HGSD%20Subsidence%20Map%201906-2000.pdf. Retrieved 1/18/2010. Last modified 1/19/2010.

Horning Geosciences. 1998. Illustration showing principal geologic hazards acting along typical sea cliff in Oregon.

ICC (International Code Council). 2012a. *International Building Code.* Birmingham, AL.

ICC. 2012b. *International Residential Code.* Birmingham, AL.

ICC. 2012c. *International Wildland-Urban Interface Code.* Birmingham, AL.

IPCC (Intergovernmental Panel on Climate Change). 2007. *Climate Change 2007: The Physical Science Basis, Contribution of Working Group I to the Fourth Assessment Report of the Intergovernmental Panel on Climate Change.* Cambridge University Press.

Jarrell, J.D., B.M. Mayfield, E.N. Rappaport, and C.W. Landsea. 2001. *The Deadliest, Costliest, and Most Intense United States Hurricanes from 1900 to 2000 (and Other Frequently Requested Hurricane Facts).* NOAA Technical Memorandum. NWS-TPC-3, 30 pp.

Jones, C. P.; D. L. Hernandez; W. C. Eiser. 1998. "Lucas vs. South Carolina Coastal Council, Revisited." *Proceedings of the 22nd Annual Conference of the Association of State Floodplain Managers.*

Kaminsky, G. M.; R. C. Daniels; R. Huxford; D. McCandless; P. Rugegiero. 1999. "Mapping Erosion Hazard Areas in Pacific County, Washington." *Coastal Erosion Mapping and Management, Journal of Coastal Research.* Special Issue No. 28, M. Crowell and S. P. Leatherman, eds.

Keillor, J. P. 1998. *Coastal Processes Manual: How to Estimate the Conditions of Risk to Coastal Property from Extreme Lake Levels, Storms, and Erosion in the Great Lakes Basin.* 2nd Edition. WISCU-H-98-003. University of Wisconsin Sea Grant Institute.

Knowles, S.; T. A. Terich. 1977. "Perception of Beach Erosion Hazards at Sandy Point, Washington." *Shore and Beach,* Vol. 45, No. 3, pp. 31–35.

Larsen, C.E. 1994. "Beaches ridges as monitors of isostatic uplift in the Upper Great Lakes." *Journal of Great Lakes Research.* Internat. Assoc. Great Lakes Res. 20(1):108-134.

NOAA (National Oceanic and Atmospheric Administration). 2010. "The Saffir-Simpson Hurricane Wind Scale." http://www.nhc.noaa.gov/pdf/sshws.pdf. Accessed December 16, 2010.

NOAA. 2011a. Atlantic Oceanographic and Meteorological Laboratory. "Hurricane Research Division: Frequently Asked Questions." http://www.aoml.noaa.gov/hrd/tcfaq/tcfaqHED.html. Version 4.4. June. Accessed June 24, 2011.

NOAA. 2011b. Center for Operational Oceanographic Products and Services. "Mean Sea Level Trend 8534720 Atlantic City, New Jersey." http://tidesandcurrents.noaa.gov/sltrends/sltrends_station. shtml?stnid=8534720. Accessed June 16, 2011.

National Research Council. 1990. *Managing Coastal Erosion.* National Academy Press.

Shepard, F. P.; H. L. Wanless. 1971. *Our Changing Coastlines.* McGraw-Hill, Inc.

Sparks, P. R.; S. D. Schiff; T. A. Reinhold. 1994. "Wind Damage to Envelopes of Houses and Consequent Insurance Losses." *Journal of Wind Engineering and Industrial Aerodynamics.* Vol. 53, pp. 145–155.

TTU (Texas Tech University).2004. *A Recommendation for an Enhanced Fujita Scale.* Wind Science and Engineering Center. June.

USACE (U.S. Army Corps of Engineers). 1971. *Report on the National Shoreline Study.*

USACE. 2002. *Ice Engineering.* Engineering Manual EM-1110-2-1612.

USACE. 2008. *Coastal Engineering Manual.* Engineering Manual EM-1110-2-1100.

USACE. 2009a. *Water Resource Policies and Authorities Incorporating Sea-Level Change Considerations in Civil Works Program.* Engineering Circular No. 1165-2-211.

USACE. 2009b. *Dynamic Sustainability: Shoreline Management on Maryland's Atlantic Coast.* Tales of the Coast. November.

Walton, T. L.; J. P. Ahrens; C. L. Truitt; R. G. Dean. 1989. *Criteria for Evaluating Coastal Flood-Protection Structures.* Technical Report CERC 89-15. U.S. Army Corps of Engineers, Coastal Engineering Research Center

Zhang, K. 1998. *Storm Activity and Sea Level Rise along the US East Coast during the 20th Century, and Their Impact on Shoreline Position.* Ph.D. Dissertation. University of Maryland at College Park.

Siting

Siting residential buildings to minimize their vulnerability to coastal hazards should be one of the most important aspects of the development (or redevelopment) process. Informed decisions regarding siting, design, and construction begin with a complete and detailed understanding of the advantages and disadvantages of potential sites for coastal construction. Gaining this knowledge *prior* to the purchase of coastal property and the initiation of design is important to ensure that coastal residential buildings are properly sited to minimize risk.

CROSS REFERENCE

For resources that augment the guidance and other information in this Manual, see the Residential Coastal Construction Web site (http://www.fema.gov/rebuild/mat/fema55.shtm).

Experience has shown that not all coastal lands are suitable for development, or at least not the type and intensity of development that has occurred on some coastal lands in the past. Prudent siting has often been overlooked or ignored in the past; properties have been developed and buildings have been constructed close to the shoreline, near bluff edges, and atop steep coastal ridges. Unfortunately, many similar siting and development decisions are still made every day based on site conditions at the time of purchase or on an incomplete or inaccurate assessment of existing and future conditions. Too often, these decisions leave property owners and local governments struggling with a number of *avoidable* problems:

- Damage to, or loss of, buildings

- Damage to attendant infrastructure

- Buildings located on public beaches as shorelines erode

- Vulnerable buildings and infrastructure that require emergency or permanent protection measures and/or relocation

NOTE

One of the principal objectives of this Manual is to improve site selection for coastal buildings.

■ Emergency evacuation

■ Injuries and loss of life

A thorough evaluation of coastal property for development purposes involves four steps (see Figure 4-1):

1. Compile lot/parcel information for one or more candidate properties; for each property, follow steps 2 through 4.

2. Identify hazards and assess risk.

3. Determine whether the risk can be reduced through siting, design, or construction and whether the residual risks to the site and the building are acceptable.

Figure 4-1.
Evaluation of coastal
property

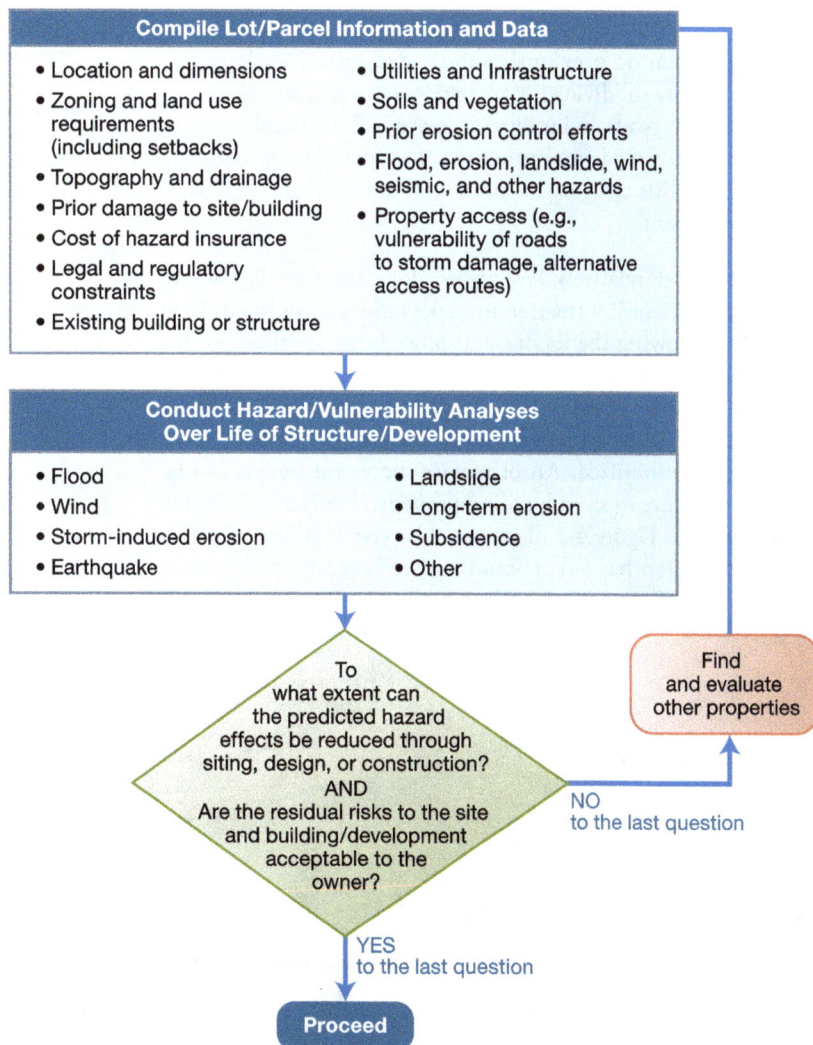

Compile Lot/Parcel Information and Data

- Location and dimensions
- Zoning and land use requirements (including setbacks)
- Topography and drainage
- Prior damage to site/building
- Cost of hazard insurance
- Legal and regulatory constraints
- Existing building or structure

- Utilities and Infrastructure
- Soils and vegetation
- Prior erosion control efforts
- Flood, erosion, landslide, wind, seismic, and other hazards
- Property access (e.g., vulnerability of roads to storm damage, alternative access routes)

Conduct Hazard/Vulnerability Analyses Over Life of Structure/Development

- Flood
- Wind
- Storm-induced erosion
- Earthquake

- Landslide
- Long-term erosion
- Subsidence
- Other

To what extent can the predicted hazard effects be reduced through siting, design, or construction? AND Are the residual risks to the site and building/development acceptable to the owner?

Find and evaluate other properties

NO to the last question

YES to the last question

Proceed

4. Either proceed with the purchase or development of a property, or reject the candidate properties, and find and evaluate other properties.

A building or development site need not be vacant or undeveloped land. Indeed, much of the construction occurring in coastal communities today involves replacement of existing buildings, infill development between adjacent buildings, or redevelopment of previously developed property (refer to Figure 4-2). This chapter addresses property evaluation broadly and applies to the following types of development:

- **Development of raw land.** Development on large, vacant parcels, usually without existing on-site access roads and utilities.

- **Development on previously subdivided lots.** Development on previously subdivided or platted lots or small parcels, usually with roads and utilities in place and surrounded by or adjacent to residential structures. Lots may or may not be vacant. This category includes infill development and redevelopment.

Today, there are relatively few places along the shoreline where there is insufficient information to make rational, informed siting decisions. Following the lessons and procedures described in this Volume of the Manual will help designers, purchasers, owners, developers, and community officials identify those locations where coastal residential development and buildings can be sited so that the risks are minimized. An otherwise successful design can be negated by failure to site a building properly. The North Carolina house shown in Figure 4-3 illustrates this type of failure; while the house appears to be a structural success, long-term erosion has left it standing permanently in the water and uninhabitable. In contrast, a siting

WARNING

Many coastal property buyers fail to investigate potential risk to their land and buildings. Designers should work with owners to identify and mitigate those risks.

WARNING

Some severe coastal hazards cannot be mitigated through design and construction. A design and construction "success" can be rendered a failure by poor siting.

WARNING

The NFIP does not insure buildings that are entirely over water or principally below ground.

Figure 4-2. Redevelopment on a previously developed lot as part of the rebuilding process after Hurricane Katrina (Lakeview, LA)

Figure 4-3.
Long-term erosion left
this well-built Kitty Hawk,
NC, house standing in the
ocean (Hurricane Dennis,
1999)
SOURCE: D. GATLEY, FEMA

Figure 4-4.
Although sited away
from the shore, winds
from Hurricane Floyd
(1999) tore off the large
overhanging roof of this
house in Wrightsville
Beach, NC

success can be overshadowed by poor design, construction, or maintenance. The North Carolina house shown in Figure 4-4 was set back from the shoreline and safe from long-term erosion, but, it could not resist winds from Hurricane Floyd in 1999.

4.1 Identifying Suitable Property for Coastal Residential Structures

The first step in the coastal development or construction process involves the purchase of a vacant or previously developed lot or parcel. This step, in many ways, constrains subsequent siting, design, and construction decisions and determines the long-term vulnerability of coastal residential buildings. *Prospective property buyers who fail to fully investigate properties before acquiring them may subsequently be faced with a variety of problems that are difficult, costly, or essentially impossible to solve.*

Although this Manual does not address the initial identification of candidate properties in detail, buyers and design professionals who assist them with property evaluations should keep the following in mind as they narrow their search for a suitable building/development site:

- The geographic region or area a buyer is interested in determines the *hazards* to which the property is exposed.

- An *existing erosion control structure* on or near a lot or parcel is an indication of prior erosion, but the structure cannot be assumed to be adequate to protect a building or development in the future.

- The *vulnerability of a coastal building generally increases with time,* as a result of one or more of the following: gradual weakening or deterioration of the building itself; sea level or lake level rise; or erosion-induced shoreline recession, which affects the majority of coastal areas in the United States.

- *Future development activities* and patterns on adjacent and nearby properties may affect the vulnerability of buildings or development on any given property.

- Any given lot or parcel *may not be suitable for the purchaser's intended use* of the property.

- Land use, zoning, setbacks, public health regulations, floodplain management, building code, and related requirements generally determine development densities, building size and location limitations, minimum design and construction practices, and allowable responses to erosion hazards; however, *compliance with these requirements does not ensure the future safety* of the building or development.

WARNING

Before any purchase, each buyer should, in consultation with experts and local officials, determine the acceptable level of residual risk and decide how to manage the actual risks expected over the life of the building or development. Note that *risk assessment, risk tolerance, and risk reduction issues are not simple*—property acquisition and development decisions should be based on a wide range of information.

CROSS REFERENCE

Refer to Chapter 3 for discussion of coastal hazards, including flooding, erosion, wind, earthquake, and other environmental considerations.

Refer to Chapter 6 for descriptions of risk assessment, risk tolerance, and residual risk.

- Development practices that perpetuate or duplicate historical siting, design, or construction practices do not ensure the future safety of new buildings and/or development. *Many historical practices are inadequate* by today's standards; further, changing shoreline conditions may render those practices obsolete.

- *Property selection*—along with subsequent siting, design, construction, and maintenance decisions—*determines the vulnerability* of and risk to any building or improvements.

Narrowing the search for coastal property suitable for development or redevelopment requires careful consideration of a variety of property and area characteristics, including the nature and success of previous erosion control efforts (e.g., groins and revetments). Note that some communities and States restrict or prohibit the construction or reconstruction of revetment, seawall, and groin structures such as those shown in Figure 4-5.

A number of States require that residential real estate transactions be accompanied by a disclosure of information pertaining to flood hazards and other hazards (if the seller or agent knows of such hazards). However, the requirements concerning the form and timing of disclosures differ. Therefore, the type and amount of information that must be disclosed varies widely. Taken collectively, the disclosure requirements

Figure 4-5.
Groins were installed
in an attempt to stop
erosion (note narrower
beaches downdrift of
groins, as shown also in
Figure 2-12)

SOURCE: BONNIE M.
BENDELL, NORTH CAROLINA
DIVISION OF COASTAL
MANAGEMENT, USED WITH
PERMISSION

(in force and as proposed) provide a good indication of the types of information that prospective property buyers and designers should seek, whether or not their State requires such disclosure. Builders should contact a real estate agent or real estate attorney for a list of real estate natural hazard disclosure laws in their State.

4.2   Compiling Information on Coastal Property

After candidate properties are identified, the next step is to compile a wide range of information for each property. This is no trivial matter; this step may require considerable time and effort. Table 4-1 is a list of general information that should be compiled. Information listed in Table 4-1 is usually available from local, regional, State, or Federal governments, from universities, or from knowledgeable professionals; however, the availability and quality of the information will vary by State and community.

Table 4-1. General Information Needed to Evaluate Coastal Property

Property Location	
• Township/county/jurisdiction • Street address • Parcel designation/tax map ID • Subdivision information	• Special zoning or land use districts • Other hazard area designation • Natural resource protection area designation

Property Dimensions

- Total acreage
- Water-ward property boundary (platted or fixed line; moving line [e.g., mean high water line, mean low water line, or other datum, elevation, feature])
- Property shape
- Property elevations and topography
- Location relative to adjacent properties
- Configuration of adjacent properties
- Shoreline frontage (i.e., dimension parallel to shoreline)
- Property depth (i.e., dimension perpendicular to shoreline)
- Acreage landward/outside of natural, physical, or regulatory construction or development limits (i.e., usable acreage)

Planning and Regulatory Information

- Hazard Mitigation Plan
- Land use designation at property and adjacent properties
- Zoning classification and resulting restrictions on use
- Building code and local amendments
- Flood hazard area: elevation and construction requirements
- Erosion hazard area: construction setbacks and regulations
- Natural resource protection area: siting, construction, or use restrictions
- Easements and rights-of-way on property (including beach access locations for nearby properties or the general public)
- Local and State siting and construction regulations
- Regulatory front, back, and side setbacks
- Local and State permitting procedures and requirements
- Local and State regulations regarding use, construction, and repair of erosion control measures
- Riparian rights
- Local and State restrictions on cumulative repairs or improvements
- Conditions or other requirements attached to building or zoning permits
- Subdivision plat covenants and other restrictions imposed by developers and homeowner's associations
- Hazard disclosure requirements for property transfer, including geologic hazard reports

Physical and Natural Characteristics

- Soils, geology, and vegetation – site and regional
- Topography of nearshore (including nearshore slope), beach, dune, bluff, uplands
- Site drainage – surface water and groundwater
- Littoral sediment supply and sediment budget
- Storm, erosion, and hazard history of property
- Erodibility of the nearshore bottom
- Erosion control structure on site – type, age, condition, and history
- Proximity to inlets and navigation structures
- Previous or planned community/regional beach/dune restoration projects
- Relative sea level/water level changes – land subsidence or uplift

Infrastructure and Supporting Development

- Access road(s)
- Emergency evacuation route(s)
- Electric, gas, water, telephone, and other utilities – onsite or offsite lines and hookups
- Sewer or septic requirements/limitations
- Limitations imposed by utility/infrastructure locations on property use

Table 4-1. General Information Needed to Evaluate Coastal Property (concluded)

Financial Considerations
• Intended use – owner-occupied or rental property
• Real estate taxes
• Development impact fees
• Permit fees
• Hazard insurance – availability, premiums, deductibles, and exclusions
• Property management fees
• Special assessments for community/association projects (e.g., private roads and facilities, dune preservation)
• Maintenance and repair of private erosion control structures
• Increased building maintenance and repairs in areas subject to high winds, wind-driven rain, and/or salt spray
• Building damage costs (insured and uninsured) from previous storms

Communities participating in the NFIP should have a FIRM and FIS on file for the community (see Section 3.6.3). The FIS includes detailed flood hazard data for parts of the community and usually includes a narrative of the flood history of a community.

The best source of current hazard information is at the local level due to the local officials' knowledge of local hazards, policies, codes, and regulations. Many States and communities produce brochures or publications to help property owners and prospective buyers evaluate coastal property. The publications listed below are examples of the types of information available.

- *Natural Hazard Considerations for Purchasing Coastal Real Estate in Hawai'i: A Practical Guide of Common Questions and Answers* (University of Hawaii Sea Grant College Program 2006), answers common questions that are considered when purchasing developed and undeveloped coastal real estate. It includes a strong focus on long-term erosion, which is the most common coastal hazard in Hawaii.

- *Living on the Coast: Protecting Investments in Shore Property on the Great Lakes* (University of Wisconsin Sea Grant Program 2004) contains a description of natural processes that affect the Great Lakes coast from glacial melt and lake level rise to local erosion. It also includes information on risk management and protecting coastal properties that is relevant to all coastal areas. The FEMA Residential Coastal Construction Web page includes a list of Web resources relevant to Great Lakes hazards adapted from the University of Wisconsin Sea Grant Program.

> **NOTE**
> Owners and prospective buyers of coastal property should contact their community or State officials for publications and data that will help them evaluate the property.

- *A Manual for Researching Historical Coastal Erosion* (Fulton 1981) describes in detail how to use historical weather data, local government records, and historical maps and photographs to understand and quantify shoreline, sea bluff, and cliff retreat. Two communities in San Diego County, CA are used as case studies to illustrate the research methods presented.

- *Questions and Answers on Purchasing Coastal Real Estate in South Carolina* (South Carolina Sea Grant Extension Program 2001) provides prospective property owners with basic information on a variety of topics, including shoreline erosion, erosion control, high winds, and hazard insurance (including earthquakes).

In the absence of current hazard information, historical records can be used to preduct future hazard conditions, impacts, and frequencies. However, natural and manmade changes at a site may render simple extrapolation of historical patterns inaccurate.

4.3   Evaluating Hazards and Potential Vulnerability

Evaluating hazards and the potential vulnerability of a building is perhaps most crucial when evaluating the suitability of coastal lands for development or redevelopment. Basing hazard and vulnerability analyses solely on building code requirements, the demarcation of hazard zones or construction setback lines, and the location and design of nearby buildings is inadequate. A recommended procedure for performing such an evaluation is outlined in the next section.

4.3.1   Define Coastal Hazards Affecting the Property

Defining the coastal hazards affecting a property under consideration for development requires close examination of both historical and current hazard information. This Manual recommends the following steps:

Step 1: Use all available information to characterize the type, severity, and frequency of hazards (e.g., flood, storm-induced and long-term erosion, accretion or burial, wind, seismic, tsunami, landslide, wildfire, and other natural hazards) that have affected or could affect the property.

Step 2: Examine the record for long-term trends (> 50–100 years), short-term trends (< 10–20 years), and periodic or cyclic variations (both spatial and temporal) in hazard events. Determine whether particularly severe storms are included in the short-term or long-term records and what effects those storms had on the overall trends. If cyclic variations are observed, determine the periods and magnitudes of the variations.

Step 3: Determine whether or not extrapolation of historical trends and hazard occurrences is reasonable. Examine the record for significant changes to the coastal system or inland and upland areas that will reduce, intensify, or modify the type, severity, and frequency of hazard occurrence at the property. The following are examples of events or processes that preclude simple extrapolation of historical trends:

> **NOTE**
>
> This Manual is intended primarily for design professionals, coastal specialists, and others with the expertise to evaluate coastal hazards and the vulnerability of sites and buildings to those hazards, and to design buildings in coastal areas. Readers not familiar with hazard and vulnerability evaluations are encouraged to seek the services of qualified professionals.

> **CROSS REFERENCE**
>
> Chapter 3 presents additional information about natural hazards in coastal areas and the effects of those hazards.
>
> Chapter 6 provides information about recurrence intervals.

- Loss of a protective dune or bluff feature that had been there for a long time may lead to increased incidence and severity of flood or erosion damage.

- Loss of protective natural habitats, such as marshes, swamps, coral reefs, and shoreline vegetation, can increase vulnerability to erosion and flooding.

- Significant increases in sea, bay, or lake levels generally increase vulnerability to flooding and coastal storm events.

- Erosion or storms may create weak points along the shoreline that are predisposed to future breaching, inlet formation, and accelerated erosion, or may expose geologic formations that are more resistant to future erosion.

- Recent or historical modifications to an inlet (e.g., construction or modification of jetties, creation or deepening of a dredged channel) may alter the supply of littoral sediments and modify historic shoreline change trends.

- Formation or closure of an inlet during a storm alters local tide, wave, current, and sediment transport patterns and may expose previously sheltered areas to damaging waves (see Figures 3-39 and 3-41 in Chapter 3).

- Widespread construction of erosion control structures may reduce the input of sediments to the littoral system and cause or increase local erosion.

- Recent seismic events may have caused uplift, settlement, submergence, or fracturing of a region, altering its hazard vulnerability to flood and other hazards.

- Changes in surface water flows, drainage patterns, or groundwater movements, and reduction in vegetative cover may increase an area's susceptibility to landslides.

- Topographic changes resulting from the retreat of a sea cliff or coastal bluff may increase wind speeds at a site.

- Exposure changes, such as the removal of trees to create future development, can increase wind pressures on existing buildings at a site.

Step 4: Forecast the type, severity, and frequency of future hazard events likely to affect the property over a suitably long period of time, say over at least 50–70 years. This forecast should be based on either: (1) extrapolation of observed historical trends, modified to take into account those factors that will cause deviations from historical trends; or (2) detailed statistical and modeling studies calibrated to reflect basic physical and meteorological processes, and local conditions. Extrapolation of trends should be possible for most coastal sites and projects. Detailed statistical and modeling studies may be beyond the scope and capabilities of many coastal development projects.

WARNING

Compliance with minimum siting requirements administered by local and State governments does not guarantee a building will be safe from hazard effects. To reduce risks from coastal hazards to an acceptable level, exceeding minimum siting requirements may be necessary.

4.3.2   Evaluate Hazard Effects on the Property

Once the type, severity, and frequency of future hazard events have been forecast, designers should use past events as an indication of the nature and severity of effects likely to occur during those forecast events. Information about past events at the site of interest and at similar sites should be considered. This historical information should be combined with knowledge about the site and local conditions to estimate future hazard effects on the site and any improvements.

Designers should consider the effects of low-frequency, rare events (e.g., major storms, extreme water levels, tsunamis, earthquakes), and multiple, successive lesser events (see Figure 4-6). For example, many of the post-storm damage assessments summarized in Chapter 2 show that the cumulative erosion and damage caused by a series of minor coastal storms can be as severe as the effects of a single, major storm.

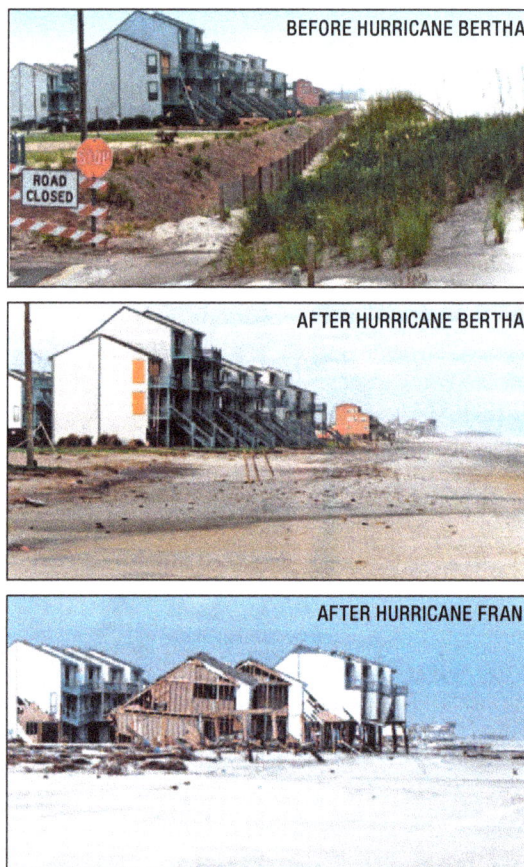

Figure 4-6.
Cumulative effects of storms occurring within a short period at one housing development in Jacksonville, NC, July–September 1996
SOURCE: JOHN ALTHOUSE, USED WITH PERMISSION

4.4   General Siting Considerations

It is always best to build in lower risk areas. However, when building in more vulnerable areas, a variety of factors must be considered in selecting a specific site and locating a building on that site. These factors are outlined in Figure 4-1 and include:

- Building code and land use requirements

- Local floodplain management requirements adopted to participate in the NFIP

- Other regulatory requirements

- Presence and location of infrastructure

- Previous development and/or subdivision of property

- Physical and natural characteristics of the property

- Vulnerability of the property to coastal hazards

When siting the foundation of a building in two different flood insurance zones, design and regulatory requirements of the most restrictive zone apply. For example, even though the majority of the foundation of the building illustrated in Figure 4-7 is located in Zone A, Zone V requirements would apply to the entire building.

Regulatory controls do not necessarily prevent imprudent siting of coastal buildings. Figure 4-8 shows flood and debris damage to new construction sited in Zone A that could have been avoided had the site been designated a Coastal A Zone, and had the structure been elevated on an open foundation. Because there are situations where minimum requirements do not address site-specific hazards, prospective buyers should

Figure 4-7.
When siting a foundation in two different flood zones, requirements for the most restrictive zone apply to the whole building

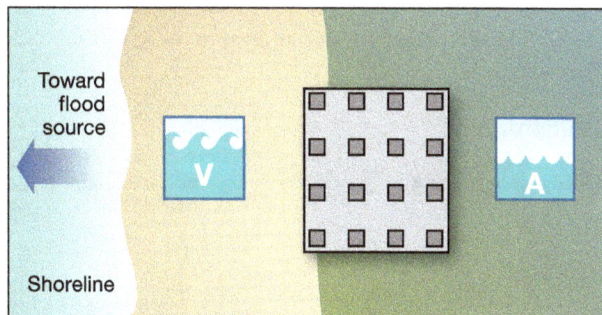

Figure 4-8.
Flood and debris damage to new construction in Zone A (Hurricane Opal, 1995)

evaluate a site for its suitability for purchase, development, or redevelopment prior to acquiring the property. However, property owners often undertake detailed studies only after property has been acquired.

Designers should recognize situations in which poor siting is allowed or encouraged, and should work with property owners to minimize risks to coastal buildings. Depending on the scale of the project, this could involve one or more of the following:

- Locating development on the least hazardous portion of the site

- Rejecting the site and finding another

- Transferring development rights to another parcel better able to accommodate development

- Combining lots or parcels

- Reducing the footprint of the proposed building and shifting the footprint away from the hazard

- Shifting the location of the building on the site by modifying or eliminating ancillary structures and development

- Seeking variances to lot line setbacks along the landward and side property lines (in the case of development along a shoreline)

- Moving roads and infrastructure

- Modifying the building design and site development to facilitate future relocation of the building on the same site

- Altering the site to reduce its vulnerability

- Construction of protective structures, if allowed by the community

> **NOTE**
>
> Proper siting and design should take into account both slow-onset hazards (e.g., long-term erosion, multiple minor storms) and rapid-onset hazards (e.g., extreme storm events).

4.5   Raw Land Development Guidelines

Large, undeveloped parcels available for coastal development generally fall into two classes:

- *Parcels well-suited to development,* but vacant due to the desires of a former owner, lack of access, or lack of demand for development. Such parcels include those with deep lots, generous setbacks, and avoidance of dune areas—these attributes should afford protection against erosion and flood events for years to come (see Figure 4-9).

- *Parcels difficult to develop,* with extensive areas of sensitive or protected resources, with topography or site conditions requiring extensive alteration, or with other special site characteristics that make development expensive relative to nearby parcels. Increasingly, coastal residential structures are planned and constructed as part of mixed-use developments, such as the marina/townhouse development shown in Figure 4-10. Such projects can involve complicated environmental and regulatory issues, as well as more difficult geotechnical conditions and increased exposure to flood hazards.

Figure 4-9.
Example of parcels
well-suited to coastal
development in Louisiana
SOURCE: USGS

Figure 4-10.
Example of parcels
difficult to develop
(mixed-use marina/
townhouse development)

Development in both circumstances should satisfy planning and site development guidelines such as those listed in Table 4-2 (adapted from recommended subdivision review procedures for coastal development in California [California Coastal Commission 1994]).

Development of raw land in coastal areas should consider the effects of all hazards known to exist and the effects of those hazards on future property owners. Similarly, such development should consider local, State, or Federal policies, regulations, or plans that will affect the abilities of future property owners to protect, transfer, or redevelop their properties (e.g., those dealing with erosion control, coastal setback lines, post-disaster redevelopment, landslides, and geologic hazards).

Table 4-2. Planning and Site Development Guidelines for Raw Land

Development of Raw Land in Coastal Areas: Summary of Site Planning and Subdivision Guidelines	
DO determine whether the parcel is suitable for subdivision or should remain a single parcel.	
DO ensure that the proposed land use is consistent with local, regional, and State planning and zoning requirements.	**DON'T** rely on engineering solutions to correct poor planning decisions.
DO ensure that all aspects of the proposed development consider and integrate topographic and natural features into the design and layout.	
DO avoid areas that require extensive grading to ensure stability.	
DO study the parcel thoroughly for all possible resource and hazard concerns.	**DON'T** assume that omissions in planning requirements can be corrected during site development.
DO identify and avoid, or set back from, all sensitive resources and prominent land features.	**DON'T** rely on relocation or restoration efforts to replace resources impacted by poor planning decisions
DO consider combining subdivision elements, such as access, utilities, and drainage.	**DON'T** overlook the effects of infrastructure location on the hazard vulnerability of building sites and lots.
DO account for all types of erosion (e.g., long-term erosion, storm-induced erosion, erosion due to inlets) and governing erosion control policies when laying out lots and infrastructure near a shoreline.	**DON'T** overlook the effects to surface and groundwater hydrology from modifications to the parcel.
DO consider existing public access to shoreline and resource areas.	**DON'T** plan development on beaches or dunes, on ridge lines or on top of prominent topographical features, on steep slopes, or in or adjacent to streams.
DO incorporate setbacks from identified high-hazard areas.	
DO use a multi-hazard approach to planning and design.	**DON'T** forget to consider future site and hazard conditions on the parcel.
DO involve a team of experts with local knowledge, and a variety of technical expertise and backgrounds.	**DON'T** assume that engineering and architectural practices can mitigate all hazards.

4.5.1 Road Placement near Shoreline

Based on studies and observations of previous coastal development patterns and resulting damage, there are several subdivision and lot layout practices that should be avoided. The first of these is placing a road close to the shoreline in an area of small lots.

In the case of an eroding shoreline, placing a road close to the shoreline and creating small lots between the road and the shoreline results in buildings, the roadway itself, and utilities being extremely vulnerable to erosion and storm damage, and can lead to future conflicts over shore protection and buildings occupying public beaches. Figure 4-11 is a view along a washed-out, shore-parallel road in Garcon Point, FL, after Hurricane Ivan in 2004. Homes

WARNING

Proper lot layout and siting of building along an eroding shoreline are critical. Failure to provide deep lots and to place roads and infrastructure well away from the shoreline ensures future conflicts over building reconstruction and shore protection.

to the left have lost inland access. Figure 4-12 shows a recommended lot layout that provides sufficient space to comply with State/local setback requirements and avoid damage to dunes. Some communities have land development regulations that help achieve this goal. For example, the Town of Nags Head, NC, modified its subdivision regulations in 1987 to require all new lots to extend from the ocean to the major shore-parallel highway (Morris 1997). Figure 4-13 compares lots permitted in Nags Head prior to 1987 with those required after 1987. The town also has policies and regulations governing the combination of nonconforming lots (Town of Nags Head 1988).

Figure 4-11.
Roads placed near shorelines can wash out, causing access problems for homes such as these located at Garcon Point, FL (Hurricane Ivan, 2004)

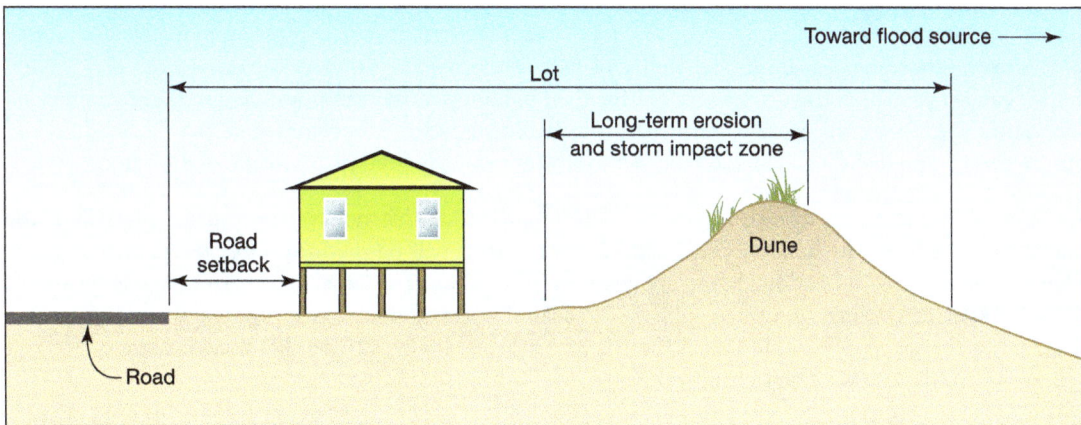

Figure 4-12.
Recommended lot layout for road setback near the shoreline

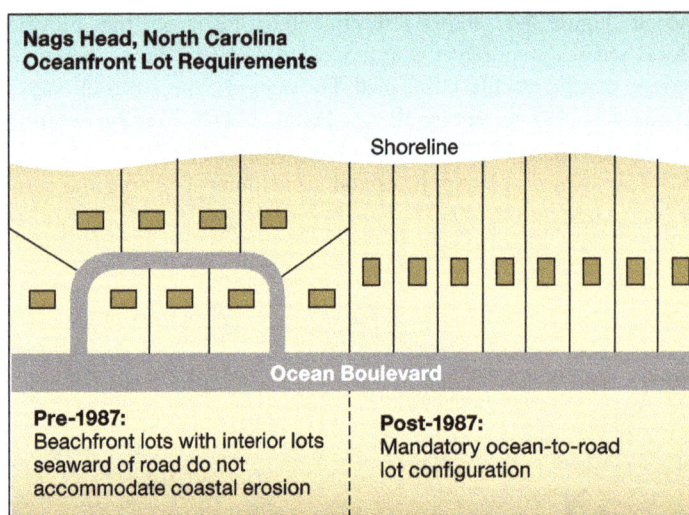

Figure 4-13.
Comparison of Nags
Head, NC, oceanfront lot
layouts permitted before
and after 1987
SOURCE: ADAPTED FROM
MORRIS 1997

A second problem associated with a shore-parallel road close to the shoreline is storm erosion damage to the road and utilities associated with the road. Some infrastructure damage can be avoided by reconfiguring the seaward lots (so they all have access from shore-perpendicular roads), eliminating the shore-parallel road, and eliminating the shore-parallel utility lines. Figure 4-14 shows shore-parallel roadways and associated utilities that may be vulnerable to storm effects and erosion (upper portion of figure). One alternative to reduce this vulnerability is to create lots and infrastructure without the shore-parallel road, and to install shutoff valves on water and sewer lines (lower portion of figure).

4.5.2   Lot Configurations along Shoreline

Another type of lot layout that is not recommended for vulnerable or eroding coastal shorelines is the "flag" lot or "key" lot illustrated in Figure 4-15. The top layout shown in the figure provides more lots with direct access to the shoreline, but limits the ability of half of the property owners to respond to coastal flood hazards and erosion by constructing or relocating their buildings farther landward. Again, the recommended alternative is to locate the shore-parallel road sufficiently landward to accommodate coastal flooding and future erosion and to create all lots so that their full width extends from the shoreline to the road.

Creation of lots along narrow sand spits and low-lying landforms is not recommended, especially if the shoreline is eroding. Any buildings constructed in such areas will be routinely subjected to coastal storm effects, overwash, and other flood hazards. Figure 4-16 shows construction along a narrow, low-lying area of Dauphin Island, AL, that is routinely subjected to coastal storm effects. Storm surge and waves transported sand across the island during Hurricane Katrina in 2005, essentially shifting the island landward. Most of the houses in this area were destroyed.

Lots should not be created in line with natural or manmade features that concentrate floodwaters (see Figure 4-17). These features can include areas of historic shoreline breaching, roads or paths across dunes, drainage features or canals, and areas of historic landslides or debris flows. Lots located landward of openings between dunes or obstructions may be more vulnerable to flooding and wave effects. Front-row lots waterward of interior drainage features may be vulnerable to concentrated flooding from the inland or bay side. One

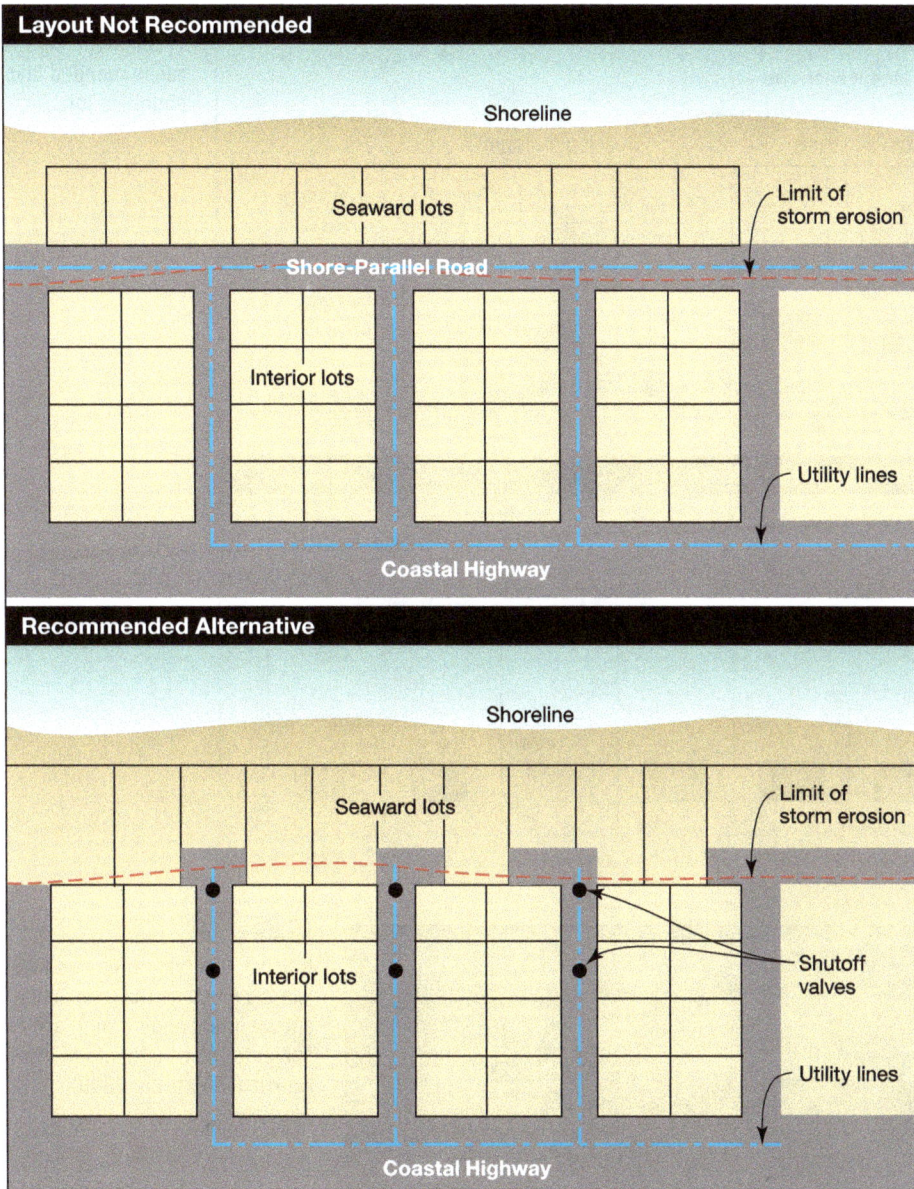

Figure 4-14.
Problematic versus recommended layouts for shore-parallel roadways and associated utilities

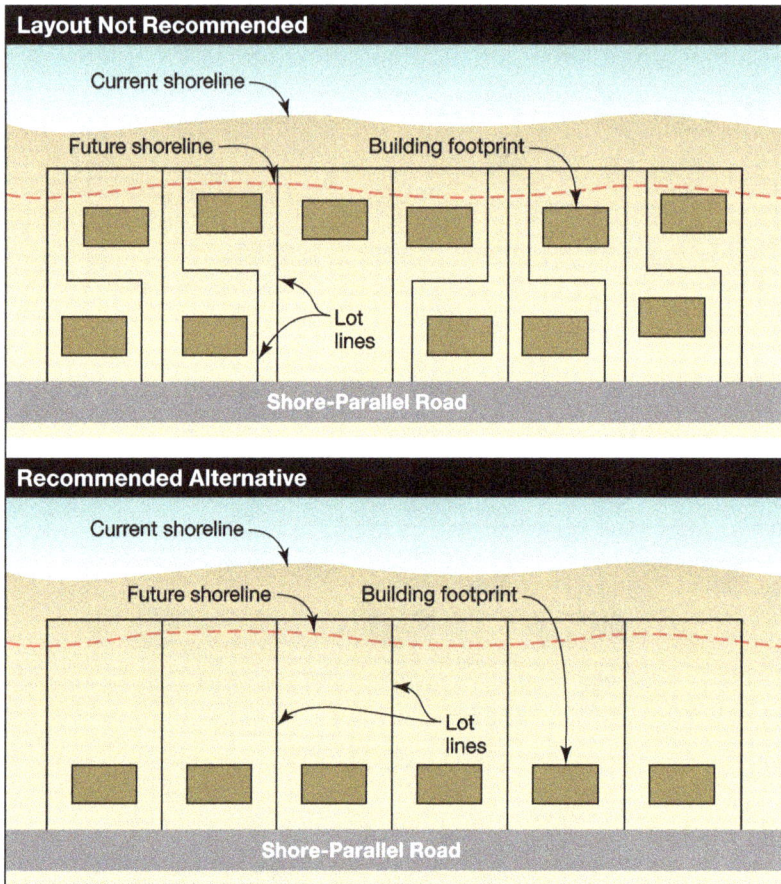

Figure 4-15.
Problematic versus recommended layouts for shoreline lots

Layout Not Recommended
- Current shoreline
- Future shoreline
- Building footprint
- Lot lines
- Shore-Parallel Road

Recommended Alternative
- Current shoreline
- Future shoreline
- Building footprint
- Lot lines
- Shore-Parallel Road

Figure 4-16.
Narrow, low-lying areas and barrier islands (such as Dauphin Island, AL, shown in the photograph) are routinely subjected to coastal storm effects (Hurricane Katrina, 2005)
SOURCE: USGS

Figure 4-17.
Lots created in line with natural or manmade features can concentrate floodwaters

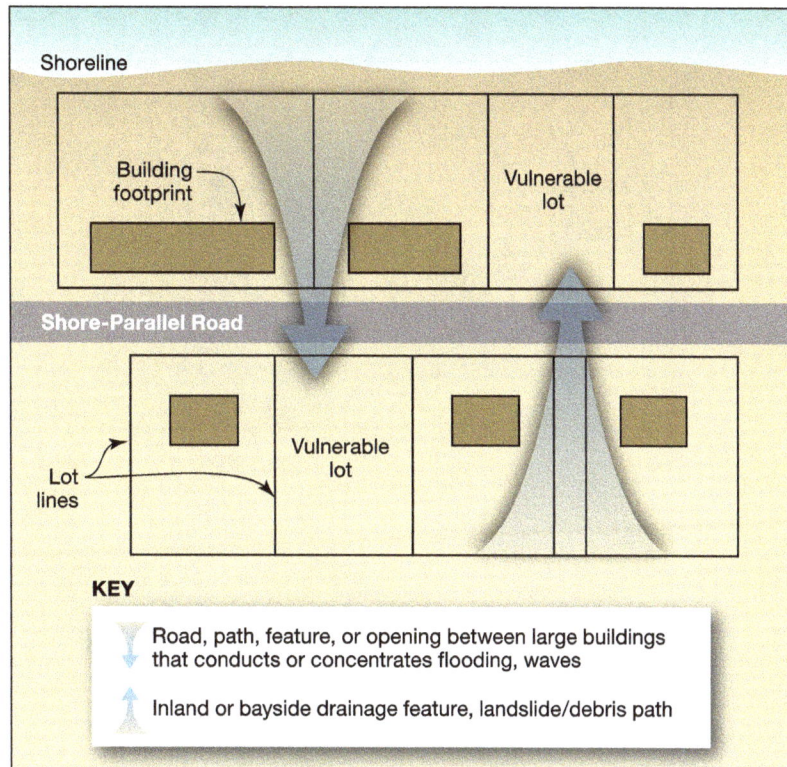

alternative is to leave these vulnerable areas as open space or to modify them to reduce associated hazards to adjacent lots. Care should also be exercised when lots are created landward of or in gaps between large buildings or objects capable of channeling floodwaters and waves (see Figures 3-20, 3-21, and 3-22).

Configurations should not concentrate small lots along an eroding or otherwise hazardous shoreline. Creating deeper lots, locating building sites farther landward on the lots, or clustering development away from the shoreline is preferable. Figure 4-18 illustrates this progression, from a "conventional" lot layout, to a "modified" lot layout, to a "cluster development" layout with lot line changes. The California Coastal Commission (1994) also developed similar alternatives for a parcel on a ridge top with steep slopes and for a parcel bisected by a coastal lagoon. Another related approach is to occupy a small fraction of the total buildable parcel and to accommodate erosion by moving threatened buildings to other available sites on the parcel. A small Pacific Ocean community in Humbolt County, CA, successfully employed this approach (Tuttle 1987), as shown in Figure 4-19, which shows a community of 76 recreational cabins on a 29-acre parcel, jointly owned by shareholders of a corporation. As buildings are threatened by erosion, they are relocated (at the building owners' expense) to other sites on the parcel, in accordance with a cabin relocation policy adopted by the corporation.

NOTE

Some States and communities have adopted regulations requiring that buildings sited in erosion-prone areas be movable. For example, Michigan has such a requirement.

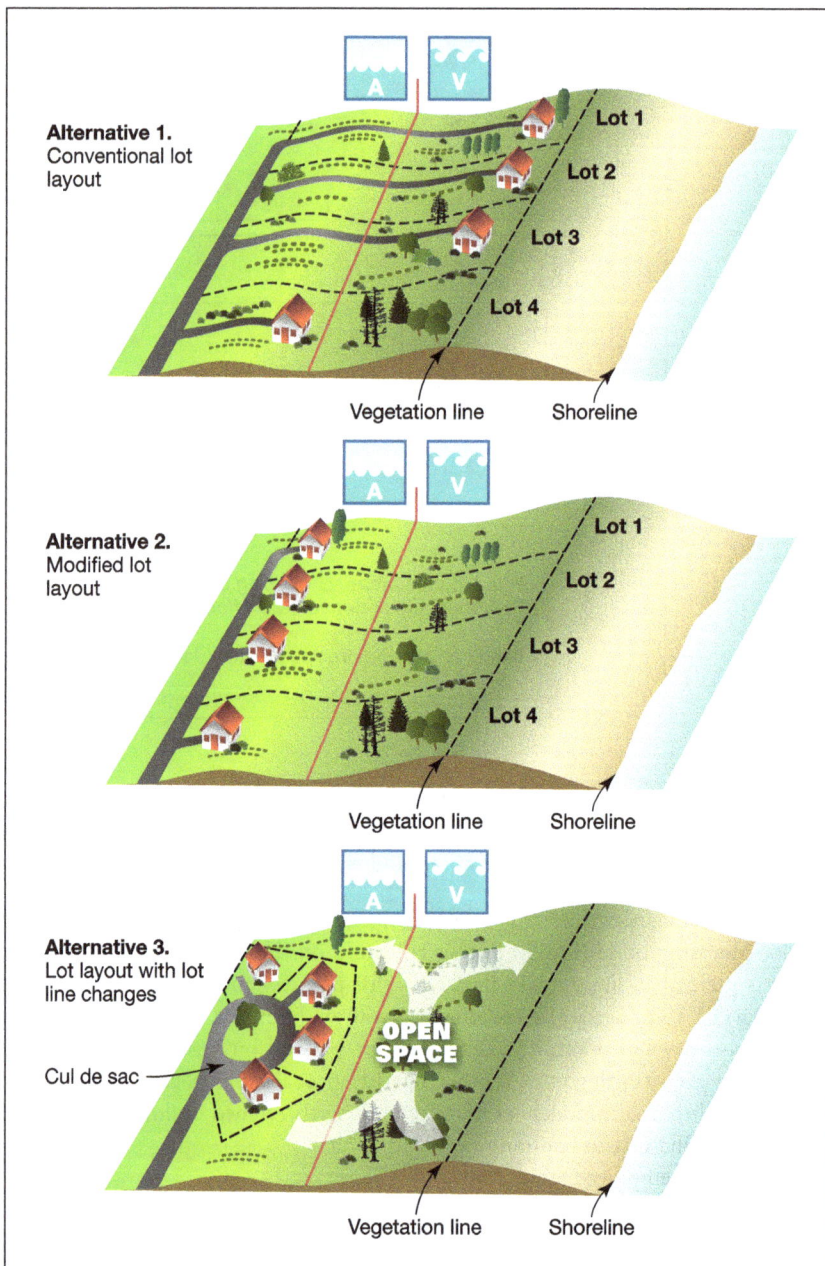

Figure 4-18.
Coastal lot development scenarios
SOURCE: ADAPTED FROM CALIFORNIA COASTAL COMMISSION 1994

Figure 4-19.
As buildings in this
Humbolt County,
CA, community are
threatened by bluff
erosion along the
Pacific Ocean, they are
moved to other sites on
the jointly owned parcel

In extreme cases, entire communities have been threatened by erosion and have elected to relocate. For example, the village of Shishmaref, AK, voted in November 1998 to relocate their community of 600 after storm erosion threatened several houses and after previous shore protection efforts failed.

More information on specific examples of relocation of threatened buildings can be found in FEMA 257, *Mitigation of Flood and Erosion Damage to Residential Buildings in Coastal Areas* (FEMA 1994). The report also presents several examples of flood and erosion mitigation through other measures (e.g., elevation, foundation alterations).

4.5.3 Lot Configurations near Tidal Inlets, Bay Entrances, and River Mouths

Layout of lots and infrastructure along shorelines near tidal inlets, bay entrances, and river mouths is especially problematic. The three South Carolina houses in Figure 4-20 were built between January 1995 and January 1996, approximately 2 years before the photograph was taken in July 1997. They were built 100 or more feet landward of the vegetation line, but rapid erosion associated with a nearby tidal inlet left the houses standing on the beach only two years after construction. The shoreline will probably return to its former location, taking several years to do so. Although the buildings are structurally intact, their siting can be considered a failure.

CROSS REFERENCE

Section 3.5 also describes instances where the subdivision and development of oceanfront parcels near ocean-bay connections led to buildings being threatened by inlet-caused erosion.

Figure 4-21 shows condominiums built adjacent to the shore in Havre de Grace, MD, where the mouth of the Susquehanna River meets the head of the Chesapeake Bay. Although the buildings are elevated, they are subject to storm surge and flood-borne debris. Infrastructure development and lot layout in similar cases should be preceded by a detailed study of historical shoreline changes, including development of (at least) a conceptual model of shoreline changes. Potential future shoreline positions should be projected, and development should be sited sufficiently landward of any areas of persistent or cyclic shoreline erosion.

Figure 4-20.
Three 2-year-old South
Carolina houses left
standing on the beach as
a result of rapid erosion
associated with a nearby
tidal inlet (July 1997)

Figure 4-21.
Condominiums built
along the shoreline
at the mouth of the
Susquehanna River on
the Chesapeake Bay were
subjected to flood-borne
debris after Hurricane
Isabel (Havre de Grace,
MD, 2003)

4.6   Development Guidelines for Existing Lots

Many of the principles discussed in the raw land scenario also apply to the construction or reconstruction of buildings on existing lots. Builders siting on a specific lot should take site dimensions, site features (e.g., topographic, drainage, soils, vegetation, sensitive resources), coastal hazards, and regulatory factors into consideration. However, several factors must be considered at the lot level; these are not a primary concern at the subdivision level:

- Buildable area limits imposed by lot-line setbacks, hazard setbacks, and sensitive resource protection requirements

- Effects of coastal hazards on lot stability

- Location and extent of supporting infrastructure, utility lines, septic tanks and drain fields, etc.

- Impervious area requirements for the lot

- Prior development of the lot

- Future building repairs, relocation, or protection

- Regulatory restrictions or requirements for on-site flood or erosion control

Although the local regulations, lot dimensions, and lot characteristics generally define the maximum allowable building footprint on a lot, designers should not assume that constructing a building to occupy the entire buildable area is a prudent siting decision. Designers should consider all the factors that can affect an owner's ability to use and maintain the building and site in the future (see Table 4-3).

Table 4-3. Guidelines for Siting Buildings on Existing Lots

Development or Redevelopment of Existing Lots in Coastal Areas: Summary of Guidelines for Siting Buildings	
DO determine whether the lot is suitable for its intended use; if not, alter the use to better suit the site or look at alternative sites.	**DON'T** assume engineering and architectural practices can mitigate poor lot layout or poor building siting.
DO study the lot thoroughly for all possible resource and hazard concerns – seek out all available information on hazards affecting the area and prior coastal hazard impacts on the lot.	**DON'T** assume that siting a new building in a previous building footprint or in line with adjacent buildings will protect the building against coastal hazards.
DO account for all types of erosion (e.g., long-term erosion, storm induced erosion, erosion due to inlets) and governing erosion control policies when selecting a lot and siting a building.	**DON'T** rely on existing (or planned) erosion or flood control structures to guarantee long-term stability of the lot.
DO avoid lots that require extensive grading to achieve a stable building footprint area.	**DON'T** overlook the constraints that site topography, infrastructure and ancillary structures (e.g., utility lines, septic tank drain fields, swimming pools), trees and sensitive resources, and adjacent development plane on site development, and (if necessary) future landward relocation of the building.
DO ensure that the proposed siting is consistent with local, regional, and state planning and zoning requirements.	**DON'T** overlook the constraints that building footprint size and location place on future work to repair, relocate or protect the building—allow for future construction equipment access and room to operate on the lot.
DO identify and avoid, or set back from, all sensitive resources.	**DON'T** overlook the effects to surface and groundwater hydrology from development of the lot.
DO consider existing public access to shoreline and resource areas.	

4.6.1   Building on Lots Close to Shoreline

Experience shows that just as developers should avoid certain subdivision development practices in hazardous coastal areas, they should also avoid certain individual lot siting and development practices. One of the most common siting errors is placing a building as close to the water as allowed by local and State regulations. Although such siting is permitted by law, it can lead to a variety of avoidable problems, including increased building vulnerability, damage to the building, and eventually encroachment onto a beach. On an eroding shoreline, this type of siting often results in the building owner being faced with one of three options: loss of the building, relocation of the building, or (if permitted) protection of the building through an erosion control measure. Alternatives to this practice include siting the building farther landward than required by minimum setbacks, and designing the building so it can be easily relocated. Siting a building farther landward also allows (in some cases) for the natural episodic cycle of dune building and storm erosion without jeopardizing the building itself. Siting a building too close to a coastal bluff edge can result in building damage or loss (see Figures 3-37 and 3-46, in Chapter 3). Keillor (1998) provides guidance regarding selecting appropriate construction setbacks for bluffs on the Great Lakes shorelines; these general concepts are applicable elsewhere.

Some sites present multiple hazards, which designers and owners may not realize without careful evaluation. Figure 4-22 shows northern California homes constructed along the Pacific shoreline at the top and bottom of a coastal bluff. These homes may be subject to several hazards, including storm waves and erosion, landslides, and earthquakes. Designers should consider all hazards and avoid them to the extent possible when siting a building.

Figure 4-22. Coastal building site in Aptos, CA, provides an example of a coastal building site subject to multiple hazards
SOURCE: CHERYL HAPKE, USGS, USED WITH PERMISSION

4.6.2 Siting near Erosion Control Structures

Siting a building too close to an erosion control structure, or failing to allow sufficient room for such a structure to be built, is another problematic siting practice. Figure 4-23 shows an example of buildings constructed near the shoreline behind a rock revetment. Although this revetment likely provided some protection to the buildings, they would have been better protected were they sited farther inland from the revetment. As shown in the figure, storm waves can easily overtop the revetment and damage the buildings.

CROSS REFERENCE

For more discussion on erosion and erosion control structures, see Section 3.5. Section 3.5.2.3 specifically discusses the effects of shore protection structures.

A related siting problem that is commonly observed along ocean shorelines as well as along bay or lake shorelines, canals, manmade islands, and marina/townhouse developments is the construction of buildings immediately adjacent to bulkheads. The bulkhead along the shoreline in front of the building in Figure 4-24 was completely destroyed from a subtropical storm. Had the building in the left of the photograph not been supported by an adequate pile foundation, it would likely have collapsed. Buildings sited close to an erosion control structure should not rely on the structure to prevent undermining. Bulkheads are rarely designed to withstand a severe coastal flood and are easily overtopped by floodwaters and waves. During severe storms, landward buildings receive little or no protection from the bulkheads. In fact, if such a bulkhead fails, the building foundation can be undermined and the building may be damaged or be a total loss.

Where buildings are constructed too close to an erosion control structure or immediately adjacent to bulkheads, it may be difficult to repair the erosion control structure in the future because of limitations on construction access and equipment operation. If erosion control structures are permitted and are employed, they should be sited far enough away from any nearby buildings to provide sufficient access to the site to complete repairs.

Figure 4-23.
Damage to buildings sited behind a rock revetment close to an eroding shoreline at Garden City Beach, SC (Hurricane Hugo, 1989)

Figure 4-24.
Beach erosion and
damage due to a
destroyed bulkhead at
Bonita Beach, FL, from a
subtropical storm
SOURCE: JUDSON HARVEY,
JUNE 1982, USED WITH
PERMISSION

4.6.3   Siting Adjacent to Large Trees

Although preservation of vegetation and landscaping are an important part of the siting process, designers should avoid siting and design practices that can lead to building damage. For example, designs that "notch" buildings and rooflines to accommodate the presence or placement of large trees should be avoided (see Figure 4-25). This siting practice may lead to avoidable damage to the roof and envelope during a high-wind event due to the unusual roof shape and additional sharp corners where wind pressure is greater.

Additionally, the potential consequences of siting a building immediately adjacent to existing large trees should be evaluated carefully. The condition and species of the existing trees should be considered. The combination of wind and rain can weaken diseased trees, causing large branches to become wind-borne debris during high-wind events. Some shallow-rooted species topple when their roots pull out of rain-saturated soils. Pine trees common to the southern United States are prone to snapping in half during high-wind events.

4.6.4   Siting of Pedestrian Access

The siting of pedestrian access between a coastal building and the shoreline often gets inadequate attention when siting decisions and plans are made. Experience shows, however, that uncontrolled access can damage coastal vegetation and landforms, providing weak points upon which storm forces act. Dune blowouts and breaches of these weak points during storms often result, and buildings landward of the weak points can be subject to increased flood, wave, erosion, or overwash effects. Several options exist for controlling pedestrian (and vehicular access) to shorelines. Guidance for the planning, layout, and construction of access structures and facilities can be found in a number of publications (additional dune walkover guidance is available on the FEMA Residential Coastal Construction Web page).

Figure 4-25.
(below) Notching the building and roofline around a tree can lead to roof and envelope damage during a high-wind event

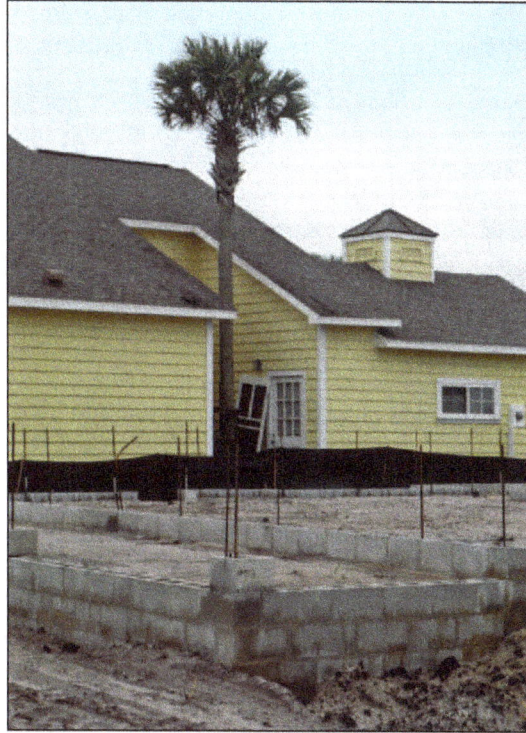

4.7   Influence of Beach Nourishment and Dune Restoration on Siting Decisions

Beach nourishment can be a means of mitigating potential adverse effects of shore protection structures. Beach nourishment and dune restoration can also be carried out alone, as a way of replacing beach or dune sediments already lost to erosion or of providing nourishment in anticipation of future erosion (National Research Council 1995).

Beach nourishment projects typically involve dredging or excavating hundreds of thousands to millions of cubic yards of sediment, and placing it along the shoreline. Beach nourishment projects are preferred over hardened erosion control structures by many States and communities, largely because the projects add sediment to the littoral system and provide recreational beach space.

The longevity of a beach nourishment project depends upon several factors: project length, project volume, native beach and borrow site sediment characteristics, background erosion rate, and the incidence and severity of storms following project implementation. Thus, most projects are designed to include an initial beach nourishment phase, followed by periodic maintenance nourishment (usually at an interval of 5 to 10 years).

WARNING

Beach nourishment and dune restoration projects are temporary. Although they can mitigate some storm and erosion effects, their presence should not be a substitute for sound siting, design, and construction practices.

The projects can provide protection against erosion and storm effects, but future protection is tied to a community's commitment to future maintenance efforts.

Beach nourishment projects are expensive and often controversial (the controversy usually arises over environmental concerns and the use of public monies to fund the projects). That controversy is beyond the scope of this Manual, but planning and construction of these projects can take years to carry out, and economic considerations usually restrict their use to densely populated shorelines. Therefore, as a general practice, designers and owners should not rely upon future beach nourishment to compensate for poor siting decisions.

As a practical matter, however, beach nourishment is the only viable option available to large, highly developed coastal communities, where both inland protection and preservation of the recreational beach are vital. Beach nourishment programs are ongoing in many of these communities and infill development and redevelopment continue landward of nourished beaches. Although nourishment programs reduce potential storm and erosion damage to inland development, they do not eliminate all damage, and sound siting, design, and construction practices must be followed.

Dune restoration projects typically involve placement of hundreds to tens of thousands of cubic yards of sediment along an existing or damaged dune. The projects can be carried out in concert with beach nourishment, or alone. Smaller projects may fill in gaps or blowouts caused by pedestrian traffic or minor storms, while large projects may reconstruct entire dune systems. Dune restoration projects are often accompanied by dune revegetation efforts in which native dune grasses or ground covers are planted to stabilize the dune against windblown erosion, and to trap additional windblown sediment.

WARNING

Although dune vegetation serves many valuable functions, such as stabilizing existing dunes and building new dunes, it is not very resistant to coastal flood and erosion forces.

The success of dune restoration and revegetation projects depends largely on the condition of the beach waterward of the dune. Property owners and designers are cautioned that the protection provided by dune restoration and revegetation projects along an eroding shoreline is short-lived—without a protective beach, high tides, high water levels, and minor storms will erode the dune and wash out most of the planted vegetation.

In some instances, new buildings have been sited such that there is not sufficient space waterward to construct and maintain a viable dune. In many instances, erosion has placed existing development in the same situation. A dune restoration project waterward of such structures will not be effective and therefore, those buildings in greatest need of protection will receive the least protection. Hence, as in the case of beach nourishment, dune restoration and revegetation should not be used as a substitute for proper siting, design, and construction practices.

4.8   Decision Time

The final step in evaluating a lot or parcel for potential development or redevelopment is to answer two questions:

1. Can the predicted risks be reduced through siting, design, and construction?

2. Are the residual risks to the site and building/development acceptable?

Unless both questions can be answered affirmatively, the property should be rejected (at least for its intended use) and other properties should be identified and evaluated. Alternatively, the intended use of the property might be modified so that it is consistent with predicted hazard effects and other constraints. Ultimately, however, reducing the long-term risks to coastal residential buildings requires comprehensive evaluation of the advantages and disadvantages of a given site based on sound siting practices as described in this chapter.

CROSS REFERENCE

Section 6.2.1 discusses reducing risk through design and construction. Chapter 6 also discussses residual risk.

4.9   References

California Coastal Commission. 1994. *Land Form Alteration Policy Guidance.*

FEMA (Federal Emergency Management Agency). 1994. *Mitigation of Flood and Erosion Damage to Residential Buildings in Coastal Areas.* FEMA 257. May.

Fulton, K. 1981. *A Manual for Researching Historical Coastal Erosion.* California Sea Grant Publication T-CSGCP-003.

Griggs, G. B. 1994. "California's Coastal Hazards." *Coastal Hazards – Perception, Susceptibility and Mitigation,* Special Issue No. 12. C. Finkl, ed. Prepared for the *Journal of Coastal Research, An International Forum for the Littoral Sciences.*

Keillor, J. P. 1998. *Coastal Processes Manual: How to Estimate the Conditions of Risk to Coastal Property from Extreme Lake Levels, Storms, and Erosion in the Great Lakes Basin.* WISCU-H-98-003. University of Wisconsin Sea Grant Institute.

Morris, M. 1997. *Subdivision Design in Flood Hazard Areas.* Planning Advisory Service Report Number 473. American Planning Association.

National Research Council. 1995. *Beach Nourishment and Protection.*

South Carolina Sea Grant Extension Program. 2001. *Questions and Answers on Purchasing Coastal Real Estate in South Carolina.* CR-003559. May.

Town of Nags Head. 1988. *Hurricane and Post-Storm Mitigation and Reconstruction Plan.*

Tuttle, D. 1987. "A Small Community's Response to Catastrophic Coastal Bluff Erosion." *Proceedings of the ASCE Specialty Conference, Coastal Zone '87.* American Society of Civil Engineers. New York.

University of Hawaii Sea Grant College Program. 2006. *Natural Hazard Considerations for Purchasing Coastal Real Estate in Hawai'i: A Practical Guide of Common Questions and Answers.* UNHI-SEAGRANT-BA-06-03. August.

University of Wisconsin Sea Grant Program. 2004. *Living on the Coast: Protecting Investments in Shore Property on the Great Lakes.*

Investigating Regulatory Requirements

States and communities throughout the United States enforce regulatory requirements that determine where and how buildings may be sited, designed, and constructed. These requirements include those associated with regulatory programs established by Federal and State statutes and locally adopted floodplain management ordinances, building codes, subdivision regulations, and other land use ordinances and laws. Applicable regulatory programs include the NFIP,

CROSS REFERENCE

For resources that augment the guidance and other information in this Manual, see the Residential Coastal Construction Web site (http://www.fema.gov/rebuild/mat/fema55.shtm).

which is intended to reduce the loss of life and damage caused by natural hazards, and programs established to protect wetlands and other wildlife habitat, which seek to minimize degradation of the environment. In addition, States and communities enforce requirements aimed specifically at the regulation of construction along the shorelines of oceans, bays, and lakes.

Federal, State, and local regulatory requirements can have a significant effect on the siting, design, construction, and cost of buildings. Therefore, designers, property owners, and builders engaged in residential construction projects in the coastal environment should conduct a thorough investigation to identify all regulations that may affect their properties and projects.

5.1 Land Use Regulations

State and local governments establish regulations governing the development and use of land within their jurisdictions. The goal of these land use regulations is generally to promote sound physical, social, and economic development. The regulations take many forms—zoning and floodplain management ordinances, subdivision regulations, utility codes, impact fees, historic preservation requirements, and environmental regulations—and they are often incorporated into and implemented under comprehensive or master plans developed by local jurisdictions in coordination with their State governments and under State statutory authority.

With land use regulations, communities can prohibit or restrict development in specified areas. They can also establish requirements for lot size, clearing and grading, and drainage, as well as the siting of buildings, floodplain management, construction of access roads, installation of utility lines, planting of vegetative cover, and other aspects of the land development and building construction processes. Land use regulations enacted and enforced by State and local governments across the country vary in content and complexity according to the needs and concerns of individual jurisdictions; therefore, it is beyond the scope of this Manual to list or describe specific regulations. However, such regulations can have a significant effect on the construction and improvement of residential and other types of buildings in both coastal and non-coastal areas. Therefore, designers, builders, and property owners must be aware of the regulations that apply to their projects.

WARNING

Designers and floodplain managers are cautioned that major natural hazard events can change shoreline locations, ground elevations, and site conditions. Information developed for the area *before* a significant event, including data shown on FIRMs and associated development regulations, may provide less-than-base flood protection *after* the event. Extreme care should be taken in siting and designing residential buildings in post-disaster situations.

The best sources of information about land use regulations are State and local planning, land management, economic development, building code, floodplain management, and community affairs officials. Professional organizations such as the American Planning Association (APA) and its State chapters are also excellent sources of information. Community officials may be interested in several APA projects and guidance publications (described on the APA Web site at http://www.planning.org):

- *Subdivision Design in Flood Hazard Areas* (Morris 1997), APA Planning Advisory Service Report Number 473. This report provides information and guidance on subdivision design appropriate for SFHAs and includes several examples of State and local subdivision requirements in coastal flood hazard areas. The report was prepared under a cooperative agreement with FEMA.

- *Growing Smart Legislative Guidebook* (APA 2002). Growing Smart is a major initiative launched by the APA in 1994 to examine statutory reform under the philosophy that there is no "one-size-fits-all" approach. The guidebook contains model planning statutes and commentary that highlight key issues in their use for State and local planning agencies. Chapter 7 of the guidebook includes a model "Natural Hazards Element" for incorporation into local government comprehensive plans.

- *Planning for Post-Disaster Recovery and Reconstruction* (Schwab et al. 1998), APA Planning Advisory Service Report Number 483/484. This report provides guidance regarding all hazards for local planners.

It includes a model ordinance for regulating hazard areas and includes case studies for five hazard scenarios (flood, hurricane, wildfire, earthquake, and tornado). The report includes the model "Natural Hazards Element" from the *Growing Smart Legislative Guidebook* that can be incorporated into local comprehensive plans. The report was prepared under a cooperative agreement with FEMA.

■ *Hazard Mitigation: Integrating Best Practices into Planning* (Schwab 2010), APA Planning Advisory Service Report Number 560. This report introduces hazard mitigation as a critical area of practice for planners. It provides guidance on how to integrate hazard mitigation strategies into planning activities and shows where hazard mitigation can fit into zoning and subdivision codes. The report was prepared by APA and supported by FEMA.

5.1.1   Coastal Barrier Resource Areas and Other Protected Areas

The CBRA of 1982 was enacted to protect vulnerable coastal barriers from development; minimize the loss of life; reduce expenditures of Federal revenues; and protect fish, wildlife, and other natural resources. This law established the Coastal Barrier Resources System (CBRS), which is managed by the U.S. Department of the Interior (DOI), U.S. Fish and Wildlife Service. The law restricts Federal expenditures and financial assistance that could encourage development of coastal barriers. The CBRA does not prohibit privately financed development; however, it does prohibit most new Federal financial assistance, including Federally offered flood insurance, in areas within the CBRS (also referred to as CBRA areas). Flood insurance may not be sold for buildings in the CBRS that were constructed or substantially improved after October 1, 1983. The financial risk of building in these areas is transferred from Federal taxpayers directly to those who choose to live in or invest in these areas.

> **NOTE**
>
> Additional information about CBRS regulations and areas included in the CBRS is available at the U.S. Fish and Wildlife Service Web site at http://www.fws.gov/habitatconservation/coastal_barrier.html.

The Coastal Barrier Improvement Act (CBIA), passed in 1991, tripled the size of the CBRS to over 1.1 million acres. The CBIA also designated *otherwise protected areas (OPA)* that include lands that are under some form of public ownership. The CBIA prohibits the issuance of flood insurance on buildings constructed or substantially improved after November 16, 1991, for the areas added to the CBRS, including OPAs. An exception is made to allow insurance for buildings located in OPAs that are used in a manner consistent with the purpose for which the area is protected. Examples include research buildings, buildings that support the operation of a wildlife refuge, and similar buildings. CBRS boundaries are shown on a series of maps produced by DOI.

OPA designations discourage development of privately owned inholdings and add a layer of Federal protection to coastal barriers already held for conservation or recreation, such as national wildlife refuges, national parks and seashores, State

> **NOTE**
>
> Any building within a CBRS area that is constructed or substantially improved after October 1, 1983, or the date of designation for areas added to the system in 1991, is not eligible for Federal flood insurance or other Federal financial assistance. The same restriction applies to substantially damaged buildings in a CBRS area that are repaired or renovated after those dates. However, all buildings within the CBRS must still comply with the NFIP siting, design, and construction requirements in their communities.

and county parks, and land owned by private groups for conservation or recreational purposes. The CBRS currently includes 271 OPAs, which add up to approximately 1.8 million acres of land and associated aquatic habitat.

FEMA shows approximate CBRS boundaries on FIRMs so that insurance agents and underwriters may determine eligibility for flood insurance coverage. Before constructing a new building, substantially improving an existing building, or repairing a substantially damaged building, the designer or property owner should review the FIRM to determine whether the property is located near or within CBRS or OPA boundaries. In situations where the FIRM does not allow for a definitive determination, the designer or property owner should request a determination from the U.S. Fish and Wildlife Service based on the DOI maps.

5.1.2   Coastal Zone Management Regulations

The CZMA of 1972 encourages adoption of coastal zone policies by U.S. coastal States in partnership with the Federal Government. CZMA regulations have been adopted by 28 of the 30 coastal States and the five island territories. For current information concerning the status of State and national CZM programs, refer to the Web site of the NOAA, National Ocean Service, Office of Ocean and Coastal Resource Management, at http://coastalmanagement.noaa.gov/programs/czm.html.

Each State's CZM program contains provisions to:

- Protect natural resources

- Manage development in high hazard areas

- Manage development to achieve quality coastal waters

- Give development priority to coastal-dependent uses

- Establish orderly processes for the siting of major facilities

- Locate new commercial and industrial development in or adjacent to existing developed areas

- Provide public access for recreation

- Redevelop urban waterfronts and ports, and preserve and restore historic, cultural, and aesthetic coastal features

- Simplify and expedite governmental decision-making actions

- Coordinate State and Federal actions

- Give adequate consideration to the views of Federal agencies

- Ensure that the public and local government have a say in coastal decision-making

- Comprehensively plan for and manage living marine resources

Coastal zone regulations vary greatly. Many States, such as Washington, Oregon, and Hawaii, provide guidelines for development while leaving the enactment of specific regulatory requirements up to county and local governments.

Most State CZM regulations control construction seaward of a defined boundary line, such as a dune or road. Many States, though not all, regulate or prohibit construction seaward of a second line based on erosion. Some of these lines are updated when new erosion mapping becomes available; lines that follow physical features such as dune lines are not fixed and "float" as the physical feature shifts over time. Examples of other types of State coastal regulations include requirements concerning the placement or prohibition of shore protection structures and the protection of dunes.

Some States not only control new construction, but also regulate renovations and repairs of substantially damaged buildings to a greater degree than required by the NFIP. These regulations help limit future damage in coastal areas by requiring that older buildings be brought up to current standards when they are renovated or repaired.

In addition to regulating the construction of buildings near the coast, many jurisdictions regulate the construction of accessory structures, roads and infrastructure, and other development-related activities.

5.2   National Flood Insurance Program

The NFIP, which is administered by FEMA, is a voluntary program with the goals of reducing the loss of life and damage caused by flooding, helping victims recover from floods, and promoting an equitable distribution of costs among those who are protected by flood insurance and the general public. The NFIP operates through a partnership between the Federal Government and individual communities such as States, counties, parishes, and incorporated cities, towns, townships, boroughs, and villages. Participation in the NFIP is voluntary. Lower cost, federally backed flood insurance is made available to property owners and renters in participating communities. In return, each community adopts and enforces a floodplain management ordinance or law that meets or exceeds the minimum requirements of the NFIP for new construction, substantial improvement of existing buildings, and repairs of substantially damaged buildings.

As part of administering the NFIP, FEMA conducts flood hazard studies and provides each community with FIRM and FIS reports, which together present flood hazard information, including the boundaries of the SFHA—the area subject to inundation by the flood that has a 1 percent chance of being equaled or exceeded in any given year—BFEs, and flood insurance zones. FEMA also provides State and local agencies with technical assistance and funding in support of flood hazard mitigation.

Unless the community as a whole practices adequate flood hazard mitigation, the potential for loss will not be reduced significantly. Discussed below is a history of the NFIP, and some components of the NFIP that allow for community-wide mitigation: FEMA flood hazard studies, minimum regulatory requirements enforced by communities participating in the NFIP, and the NFIP CRS program.

> **TERMINOLOGY**
>
> **SUBSTANTIAL IMPROVEMENT:**
> Improvement of a building (such as reconstruction, rehabilitation, or addition) is considered a substantial improvement if its cost equals or exceeds 50 percent of the market value of the building before the start of construction of the improvement.
>
> **SUBSTANTIAL DAMAGE:**
> Damage to a building (regardless of the cause) is considered substantial damage if the cost of restoring the building to its before-damage condition would equal or exceed 50 percent of the market value of the structure before the damage occurred.

5.2.1   History of the NFIP

Congress created the NFIP in 1968 when it passed the National Flood Insurance Act. The primary purposes of the Act are to:

- Indemnify individuals for flood losses through insurance

- Reduce future flood losses through floodplain management regulations

- Reduce Federal expenditures for disaster assistance and flood control

CROSS REFERENCE

For additional information on the NFIP and its mapping products, see Section 3.6.

FEMA is prohibited from providing flood insurance to a community under the 1968 Act if a community does not adopt and enforce floodplain management regulations that meet or exceed the floodplain management criteria established in accordance with Section 1361(c) of the 1968 Act.

Subsidizing flood insurance for existing buildings was not incentive enough for communities to voluntarily participate in the NFIP. The same held true for individuals purchasing flood insurance. In 1973, Congress passed the Flood Disaster Protection Act. The 1973 Act prohibits Federal agencies from providing financial assistance for acquisition or construction of buildings in a SFHA in a community that does not participate in the NFIP. Certain disaster assistance for these non-participating communities is also prohibited. Another key provision of the 1973 Act was the "Mandatory Flood Insurance Purchase Requirement," which requires federally insured or regulated lenders to require flood insurance on all grants and loans for buildings purchased or constructed in the SFHA.

To further the efforts of the NFIP, Congress amended the 1968 and 1973 Acts with the National Flood Insurance Reform Act in 1994. The 1994 Act: (1) increased the amount of flood insurance coverage allowed to be purchased, (2) codified the NFIP CRS, (3) added the Increased Cost of Compliance coverage for individual property owners who had to comply with floodplain management regulations, (4) established the Flood Mitigation Assistance grant program to assist States and communities to develop mitigation plans and implement measures to reduce future flood damage to structures, and (5) added a requirement that FEMA assess its flood hazard map inventory at least once every 5 years. Congress amended the 1994 Act with the Flood Insurance Reform Act of 2004. The 2004 Act established the Repetitive Flood Claims and Severe Repetitive Loss grant programs to reduce or eliminate future losses to properties in the NFIP.

5.2.2   FEMA Flood Hazard Studies

To provide communities with the information needed to enact and enforce floodplain management ordinances or laws consistent with the requirements of the NFIP, FEMA conducts flood hazard studies for communities throughout the United States and publishes the results in FIRMs and FIS reports.

CROSS REFERENCE

For an explanation of how BFEs, flood zones, and LiMWAs are determined for coastal flood hazard areas and how they affect coastal construction, see Section 3.6.

The information provided by FIS reports and FIRMs includes the names and locations of flooding sources; the sizes and frequencies of past floods; the limits of the SFHA in areas subject to riverine, lacustrine, and coastal flooding; flood insurance zone designations; and BFE contours throughout the SFHA. FIRMs in coastal areas may also show the LiMWA. Communities can use the information provided in FIS reports and FIRMs to manage SFHA development. At the same time, FEMA uses the FIS and FIRMs to establish insurance premiums for houses and other buildings. The information pertaining to the BFE and the flood zone at the building site are of particular importance for a coastal construction project.

5.2.3   Minimum Regulatory Requirements

The floodplain management ordinances or laws adopted by communities that participate in the NFIP must meet or exceed the minimum NFIP regulatory requirements set forth at Title 44 of the Code of Federal Regulations (CFR) Section 60.3 (44 CFR § 60.3). Community floodplain management regulations include requirements in the SFHA that apply to new construction, substantially improved buildings, and substantially damaged buildings in both Zone A and Zone V. Additional requirements apply to new subdivisions and other development in the SFHA.

The minimum NFIP requirements for new construction, substantially improved, and substantially damaged buildings affect the type of foundation that can be used, establishes the required height of the lowest floor to or above the BFE, establishes the criteria for the installation of building utility systems, requires the use of flood damage-resistant materials, and limits the use of the area below the lowest floor. In recognition of the greater hazard posed by breaking waves 3 feet high or higher, FEMA has established minimum NFIP regulatory requirements for Zone V buildings that are more stringent than the minimum requirements for Zone A buildings. Therefore, the location of a building in relation to the Zone A/Zone V boundary on a FIRM can affect the design of the building. In that regard, it is important to note that if a building or other structure has any portion of its foundation in Zone V, it must be built to comply with Zone V requirements.

The following sections summarize the minimum NFIP requirements (for the exact wording of the regulations, refer to 44 CFR § 60.3): Section 5.2.3.1 describes the minimum requirements that apply throughout the SFHA. Sections 5.2.3.2 and 5.2.3.3 describe requirements specific to Zone A and Zone V, respectively.

WARNING

Communities participating in the NFIP are encouraged to adopt and enforce floodplain management ordinances or laws more stringent than the minimum requirements of the NFIP regulations. For example, some States and communities require that buildings be elevated above rather than simply to the BFE. The additional elevation is referred to as freeboard (see Figure 5-4). Check with local floodplain managers and building officials concerning such requirements.

WARNING

The guidance in this Manual was not specifically developed for manufactured housing. For NFIP requirements concerning manufactured housing, refer to 44 CFR Section 60.3 and FEMA P-85, *Protecting Manufactured Homes from Flood and Other Hazards, A Multi-Hazard Foundation and Installation Guide* (FEMA 2009a).

5.2.3.1   Minimum Requirements in All SFHAs

The minimum NFIP floodplain management requirements for all SFHAs affect buildings, subdivisions and other new development, new and replacement water supply systems, and new and replacement sanitary sewage systems. These requirements, set forth at 44 CFR § 60.3(a) and (b), are summarized in Table 5-1.

Table 5-1. General NFIP Requirements

Activity	General NFIP Requirement in All SFHAs
New Construction, Substantial Improvement, and Repair of Substantially Damaged Buildings	• Communities shall require permits for development in SFHAs and shall review permit applications to determine whether proposed building sites will be reasonably safe from flooding. • Buildings shall be designed (or modified) and anchored to prevent flotation, collapse, and lateral movement resulting from hydrodynamic and hydrostatic loads, including the effects of buoyancy. • Buildings shall be constructed with materials resistant to flood damage. • Buildings shall be constructed with methods and practices that minimize flood damage. • Buildings shall be constructed with electrical, heating, ventilation, plumbing, and air conditioning equipment and other service facilities that are designed and/or located to prevent water from entering or accumulating within their components during flooding. • Communities shall obtain and reasonably use any BFE and floodway data available from other sources for SFHAs for which the FIRM does not provide BFEs or floodways.
New Subdivisions and Other New Developments	• Communities shall review proposals for subdivisions and other new developments to determine whether such proposals will be consistent with the need to minimize flood damage within flood-prone area. • Proposals for new subdivisions and other new developments greater than 50 lots or 5 acres, whichever is less, and for which BFEs are not shown on the effective FIRM shall include BFE data. • Public utilities and facilities, such as sewer, gas, electrical, and water systems for new subdivisions and other new developments shall be located and constructed to minimize or eliminate flood damage. • Adequate drainage shall be provided for new subdivisions and new developments to reduce exposure to flood hazards.
New and Replacement Water Supply Systems	• New and replacement water supply systems shall be designed to minimize or eliminate infiltration of flood waters into the systems.
New and Replacement Sanitary Sewage Systems	• New and replacement sanitary sewage systems shall be designed to minimize or eliminate infiltration of flood waters into the systems and discharges from the systems into flood waters. • On-site waste disposal systems shall be located to avoid impairment to them or contamination from them during flooding.

Floodplain management regulations apply to new construction, substantially improved buildings, and substantially damaged buildings located within the SFHA. FEMA has two resources to assist State and local officials with NFIP requirements: FEMA P-758, *Substantial Improvement/Substantial Damage (SI/SD) Desk Reference* (FEMA 2010a) and the FEMA P-784 *Substantial Damage Estimator (SDE)* software (FEMA 2010b). FEMA P-758 is intended to be used by local officials responsible for administering local codes and ordinances, including requirements related to substantial improvement and substantial damage. It also is intended for State officials who provide NFIP technical assistance to communities. FEMA P-758 provides practical guidance and suggested procedures to implement the NFIP requirements for substantial improvement and repair of substantial damage.

The SDE software was developed to assist State and local officials in determining substantial damage in accordance with a local floodplain management ordinance meeting the requirements of the NFIP. Data collected during the evaluation process and entered into the SDE software provides an inventory of potentially substantially damaged buildings, including both residential and non-residential structures. For more information, consult the local floodplain management official in the area where the building is being constructed. FEMA 213, *Answers to Questions About Substantially Damaged Buildings* (FEMA 1991; currently being updated as of the publication of this Manual) provides answers to commonly asked questions about substantial improvement and substantial damage.

> **WARNING**
>
> In addition to the floodplain management requirements discussed in this Manual, the NFIP regulations include requirements specific to floodplains along rivers and streams. Because this Manual focuses on the construction of residential buildings in coastal areas, it does not discuss these additional requirements. For more information about these requirements, consult local floodplain management officials. Also refer to FEMA 259, *Engineering Principles and Practices for Retrofitting Flood-Prone Residential Structures* (FEMA 2011).

5.2.3.2   Additional Minimum Requirements for Buildings in Zone A

The additional minimum requirements specific to buildings in Zones AE, A1–A30, AO, and A pertain to the elevation of the lowest floor, including basement, in relation to the BFE or the depth of the base flood, and to the enclosed areas below the lowest floor. Note that these requirements are the same for Coastal A Zones and Zone A.

Building Elevation in Zones AE and A1–A30

The top of the lowest floor, including the basement floor, of all new construction, substantially improved, and substantially damaged buildings must be at or above the BFE.

The lowest floors of buildings in Zones AE, A1–A30, and A must be at or above the BFE. Foundation walls below the BFE must have openings that allow the entry of flood waters so that interior and exterior hydrostatic pressures can equalize. Note that some damage is likely to be sustained if building construction meets only the minimum NFIP requirements because the structure under the top of the lowest floor will be inundated during the base flood.

Building Elevation in Zone A

FIRMs do not show BFEs in SFHAs designated Zone A (i.e., unnumbered Zone A) because detailed flood hazard studies in those areas have not been performed. The lowest floors of buildings in Zone A must be elevated to or above the BFE whenever BFE data are available from other sources. The IBC and IRC both authorize the local official to require an applicant to use BFE data from other sources or to determine the BFE. If no BFE data are available, communities must ensure that buildings are constructed with methods and practices that minimize flood damage.

Building Elevation in Zone AO

Zone AO designates areas where flooding is characterized by shallow depths (averaging 1–3 feet) and/or unpredictable flow paths. In Zone AO, the top of the lowest floor, including the basement floor, of all new construction, substantially improved, and substantially damaged buildings must be above the highest grade adjacent to the building by at least the depth of flooding in feet shown on the FIRM. For example, if the flood depth shown on the FIRM is 3 feet, the top of the lowest floor must be at least 3 feet above the highest grade adjacent to the building. If no depth is shown on the FIRM, the minimum required height above the highest adjacent grade is 2 feet.

Enclosures Below the Lowest Floor in Zones AE, A1–A30, AO, and A

Enclosed space below the lowest floors of new construction, substantially improved, and substantially damaged buildings may be used only for parking of vehicles, access to the building, or storage. The walls of such areas must have openings designed to allow the automatic entry and exit of flood waters so that interior and exterior hydrostatic pressures equalize during flood events. To satisfy this requirement, non-engineered openings may be used to provide a total net open area of 1 square inch per square foot of enclosure. Designs for engineered openings must be certified by a registered professional engineer or architect as providing the required performance (see Section 2.6.2 of ASCE 24, *Flood Resistant Design and Construction*). The installation of openings must meet the following ASCE 24 criteria:

> **WARNING**
>
> Even waves less than 3 feet high can impose large loads on foundation walls. This Manual recommends that buildings in the Coastal A Zone be designed and constructed to meet Zone V requirements (see Section 5.4.2 and Chapter 11).

1. Each enclosed area must have openings.

2. There must be a minimum of two openings on different sides of each enclosed area, and

3. The bottom of each opening must be no more than 1 foot above the higher of the final interior grade or floor and the finished exterior grade immediately under each opening.

For more information about openings requirements for the walls of enclosures below the lowest floors of buildings in Zone A, refer to FEMA NFIP Technical Bulletin 1, *Openings in Foundation Walls and Walls of Enclosures Below Elevated Buildings in Special Flood Hazard Areas in accordance with the National Flood Insurance Program* (FEMA 2008d).

> **WARNING**
>
> Flood vents must be unobstructed in order to perform as intended. For example, flood vents backed with interior gypsum board finish do not allow for the automatic entry and exit of flood waters.

5.2.3.3   Additional Minimum Requirements for Buildings in Zone V

The additional minimum requirements enforced by participating communities regarding new construction, substantially improved buildings, and substantially damaged buildings in Zones VE, V1–V30, and V pertain to the siting of the building, the elevation of the lowest floor in relation to the BFE, the foundation design, enclosures below the lowest floor, and alterations of sand dunes and mangrove stands (refer to 44 CFR § 60.3(e)).

Siting

All new construction must be located landward of the reach of mean high tide (i.e., the mean high water line). In addition, manmade alterations of sand dunes or mangrove stands are prohibited if those alterations would increase potential flood damage. Removing sand or vegetation from, or otherwise altering, a sand dune or removing mangroves may increase potential flood damage; therefore, such actions must not be carried out without the prior study and approval from a local floodplain official.

Building Elevation

All new construction, substantially improved, and substantially damaged buildings must be elevated on pilings, posts, piers, or columns so that the bottom of the lowest horizontal structural member of the lowest floor (excluding the vertical foundation members) is at or above the BFE. In Zone V, buildings must be elevated on an open foundation (e.g., pilings, posts, piers, or columns).

Foundation Design

The piling or column foundations for all new construction, substantially improved, and substantially damaged buildings, as well as the buildings attached to the foundations, must be anchored to resist flotation, collapse, and lateral movement due to the effects of wind and water loads acting simultaneously on all components of the building. A registered engineer or architect must develop or review the structural design, construction specifications, and plans for construction and must certify that the design and methods of construction to be used are in accordance with accepted standards of practice for meeting the building elevation and foundation design standards described above.

In addition, erosion control structures and other structures such as bulkheads, seawalls, and retaining walls may not be attached to the building or its foundation.

CROSS REFERENCE

For more information about enclosures, the use of space below the lowest floor, and breakaway walls, refer to Section 8.5.8, 8.5.10, 12.4, and 13.1.10 of this Manual and to the following FEMA NFIP Technical Bulletins:

- *Design and Construction Guidance for Breakaway Walls Below Elevated Buildings Located in Coastal High Hazard Areas in accordance with the National Flood Insurance Program,* Technical Bulletin 9 (FEMA 2008a)

- *Flood Damage-Resistant Materials Requirements for Buildings Located in Special Flood Hazard Areas in accordance with the National Flood Insurance Program,* Technical Bulletin 2 (FEMA 2008b)

- *Free-of-Obstruction Requirements for Buildings Located in Coastal High Hazard Areas in accordance with the National Flood Insurance Program,* Technical Bulletin 5 (FEMA 2008c)

NOTE

For more information about the use of fill in Zone V, refer to *Free-of-Obstruction Requirements for Buildings Located in Coastal High Hazard Areas in accordance with the National Flood Insurance Program,* FEMA NFIP Technical Bulletin 5 (FEMA 2008c).

Use of Fill

Fill may not be used for the structural support of any building within Zones VE, V1–V30, and V. Minor grading and the placement of minor quantities of fill is permitted for landscaping and drainage purposes under and around buildings and for support of parking slabs, pool decks, patios and walkways. Fill may be used in Zone V for minor landscaping and site drainage purposes (consult local officials for specific guidance or requirements).

Space Below the BFE

The space below all new construction, substantially improved, and substantially damaged buildings must either be free of obstructions or enclosed only by non-supporting breakaway walls, open wood latticework, or insect screening intended to collapse under water loads without causing collapse, displacement, or other structural damage to the elevated portion of the building or the supporting foundation system. Furthermore, NFIP requirements specify permitted uses below the BFE, use of flood damage-resistant materials below

SUBSTANTIAL IMPROVEMENT AND SUBSTANTIAL DAMAGE

Designers working on existing buildings should check with local officials early in the design process to find out if the proposed work is likely to trigger substantial improvement requirements. Local officials must review proposals to improve structures that are located in mapped SFHAs to determine whether the proposed work will be considered substantial improvement or repair of substantial damage.

The determination is based on comparing the cost of the improvement (or cost to repair a damaged building to its pre-damage condition) to the market value of the building before the improvement (or before the damage occurred). If the cost equals or exceeds 50 percent of the market value, the building must be brought into compliance with NFIP requirements based on its location in the flood zone and its occupancy.

The requirements apply to buildings in all SFHAs. The requirements that apply in Zone V (and those recommended for Coastal A Zones) require that substantially improved and substantially damaged buildings:

- Be elevated on open foundations (pilings or columns)
- Be elevated so that the bottom of the lowest horizontal structural member of the lowest floor is at or above the BFE
- Have the foundation anchored to resist flotation, collapse, and lateral movement due to the effects of wind and water loads acting simultaneously on all building components
- Have the area beneath the elevated building free of obstructions
- Have utility and building service equipment elevated above the BFE
- Have the walls of enclosures below the elevated building designed to break away under base flood conditions without transferring loads to the foundation
- Use flood damage-resistant materials below the BFE

Work on a post-FIRM building cannot be allowed if it would make the building noncompliant with the requirements in place at the time the building was originally constructed.

If a property owner decides to demolish an existing building and rebuild on the same site, the work is considered new construction and all requirements for new construction must be met.

SUBSTANTIAL IMPROVEMENT AND SUBSTANTIAL DAMAGE (concluded)

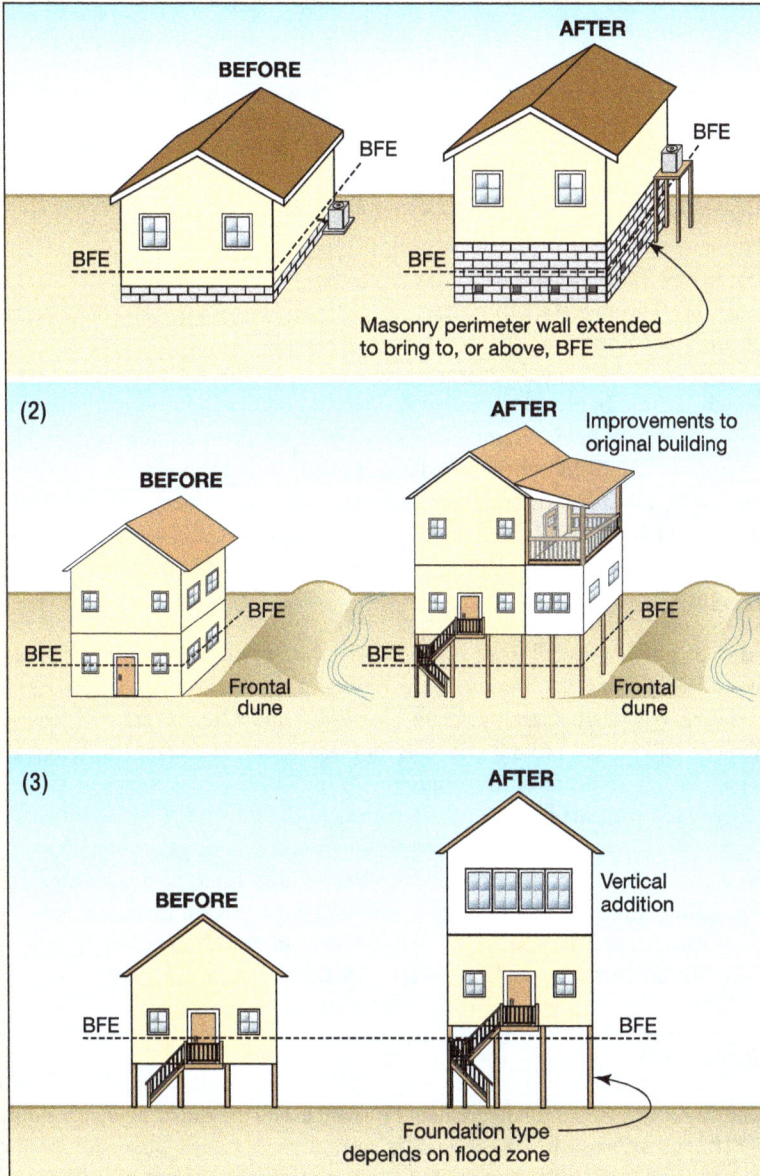

Figure above (top to bottom): Substantial improvement triggered by (1) rehabilitation with no increase in footprint to a home in Zone A (top)—building must be brought into compliance with the NFIP, (2) lateral addition to a home in Zone V (middle)—both the addition and the original building must be brought into compliance with the NFIP, and (3) vertical addition (either new upper or lower floor, bottom figure)—in this case, the whole building must be brought into compliance with the NFIP.

SOURCE: FEMA P-758 (2010a)

the BFE (see NFIP Technical Bulletin 2, FEMA 2008b), and placement of mechanical/utility equipment below the BFE. Compliance with these requirements for the space below the BFE will minimize flood damage. This has been confirmed by post-damage assessments of buildings following disaster events. Failure to comply with these requirements violates the local floodplain management ordinance and NFIP regulations, and can lead to higher flood insurance premiums and uninsured losses.

> **WARNING**
>
> Although the NFIP regulations permit below-BFE enclosures that meet the criteria presented here, many communities may have adopted ordinances that prohibit all such enclosures or that establish more stringent criteria, such as an enclosure size limitation. Check with local officials about such requirements.

The current NFIP regulatory requirements regarding breakaway walls are set forth at 44 CFR § 60.3(e)(5). The regulations specify a design safe loading resistance for breakaway walls of not less than 10 pounds per square foot and not more than 20 pounds per square foot. However, the regulations also provide guidance for the use of alternative designs that do not meet the specified loading requirements. In general, breakaway walls built according to such designs are permitted if a registered engineer or architect certifies that the walls will collapse under a water load less than that of the base flood and that the elevated portion of the building and supporting foundation system will not be subject to collapse, displacement, or other structural damage due to the simultaneous effects of wind and water loads on all components of the building. Additional requirements apply to the use of an enclosed area below the lowest floor—it may be used only for parking, building access, or storage and it must be constructed of flood damage-resistant materials.

The current NFIP regulations do not provide specifications or other detailed guidance for the design and construction of alternative types of breakaway walls. However, the results of research conducted for FEMA and the National Science Foundation by North Carolina State University and Oregon State University, including full-scale tests of breakaway wall panels, provide the basis for prescriptive criteria for the design and construction of breakaway wall panels that do not meet the requirement for a loading resistance of 10 to 20 pounds per square foot. These criteria are presented in the NFIP Technical Bulletin 9 (FEMA 2008a). The criteria address breakaway wall construction materials, including wood framing, light-gauge steel framing, and masonry; attachment of the walls to floors and foundation members; utility lines; wall coverings such as interior and exterior sheathing, siding, and stucco; and other design and construction issues. In addition, the bulletin describes the results of the testing. The test results are described in greater detail in *Behavior of Breakaway Walls Subjected to Wave Forces: Analytical and Experimental Studies* (Tung et al. 1999).

5.2.4 Community Rating System

Although a participating community's floodplain management ordinance or law must, at a minimum, meet the requirements of the NFIP regulations, FEMA encourages communities to establish additional or more stringent requirements as they see fit. In 1990, to provide incentives for communities to adopt more stringent requirements, FEMA established the NFIP CRS, a program through which FEMA encourages and recognizes community floodplain management activities that exceed the minimum NFIP requirements. Under the CRS, flood insurance premiums within participating communities are adjusted to reflect the reduced flood risk resulting from community activities that meet the three goals of the CRS: (1) reducing flood losses, (2) facilitating accurate insurance ratings, and (3) promoting awareness of the importance of flood insurance.

Through the CRS, a community is awarded credit points for carrying out floodplain management activities in the areas of public information, mapping and regulations, flood damage reduction, and flood preparedness. The number of points awarded determines the community's CRS class (from 1 to 10), which, in turn, determines the community's discount in flood insurance premiums for structures within and outside the SFHA. Participation in the CRS is voluntary; any community compliant with the rules and regulations of the NFIP may apply for a CRS classification. In addition to helping communities obtain insurance premium discounts, the CRS promotes floodplain management activities that help save lives, reduce property damage, and promote sustainable, more livable communities.

5.3   Building Codes and Standards

Many States and communities regulate the construction of buildings by adopting and enforcing building codes. Building codes set forth minimum requirements for structural design, materials, fire safety, exits, natural hazard mitigation, sanitary facilities, light and ventilation, environmental control, fire protection, and energy conservation. The purpose of a code is to establish the minimum acceptable requirements necessary for protecting the public health, safety, and welfare in the built environment. Building codes apply primarily to new construction, but may also apply to existing buildings that are being repaired, altered, or added to and when a building is undergoing a change of occupancy as defined by the code.

Numerous standards related to design and construction practices and construction materials are incorporated into a building code by reference rather than by inclusion of all of the text of the standard in the code. For example, ASCE 7 is a reference standard for both the IBC and IRC, where applicable provisions of ASCE 7 are enacted by reference, in lieu of directly incorporating text of ASCE 7 into the IBC and IRC.

Most locally adopted building codes in the United States are based on model building codes. Examples of model building codes are the series of codes promulgated by the International Code Council (ICC) including:

- *International Building Code* (IBC), (ICC 2012a)

- *International Residential Code for One- and Two-Family Dwellings* (IRC), (ICC 2012b)

NOTE

As of May 1, 2010, 1,138 communities throughout the United States were receiving flood insurance premium discounts through the CRS as a result of implementing local mitigation, outreach, and educational activities that exceed the minimum NFIP requirements. For more information about the CRS, contact the State NFIP Coordinating Agency or the appropriate FEMA Regional Office (listed on the FEMA Residential Coastal Construction Web page).

NOTE

The adoption and enforcement of building codes and standards is not consistent across the United States. Codes and standards in some States and communities may be more restrictive than those in others. In addition, some communities have not adopted a building code. In communities where building codes have not been adopted or where the existing codes are not applied to one- and two-family residential buildings, design professionals, contractors, and others engaged in the design and construction of coastal residential buildings are encouraged to follow the requirements of a model building code and the best practices presented in this Manual.

- *International Existing Building Code* (IEBC) (ICC 2012c)

- *International Mechanical Code* (IMC) (2012d)

- *International Plumbing Code* (IPC) (2012e)

- *International Private Sewage Disposal Code* (IPSDC) (2012f)

- *International Fuel Gas Code* (IFGC) (2012g)

- *International Fire Code* (IFC) (2012h)

Provisions of the IBC and IRC are the model building codes of most interest for this Manual because they address primary requirements for design and construction of coastal residential buildings and because of their wide-spread use in the United States. The National Fire Protection Association's NFPA 5000 (NFPA 2012), *Building Construction and Safety Code,* is used by some jurisdictions instead of the IBC and IRC.

While model codes are widely used, States and local jurisdictions often incorporate amendments and revisions to meet specific needs. Variations in code provisions from one State or jurisdiction to the next, coupled with potential code revisions, make it imperative that the designer work with local officials to identify applicable codes, standards, and construction requirements.

NOTE

When the 2000 I-Codes were first published, many components of the NFIP were not included. After freeboard requirements were added to the 2006 I-Codes, NFIP requirements were represented in the minimum requirements of building codes. By referencing ASCE 24, the I-Codes include some requirements more restrictive than the NFIP.

NOTE

Provisions of the IBC, IRC, IMC, IPC, IPSDC, IFGC, IFC and NFPA 5000 are consistent with applicable provisions of NFIP regulations.

Even in cases where amendments are minimal and where the commonly used model codes are adopted, questions often arise regarding the applicability of IBC and IRC code provisions to the design of residential buildings. As stated in the scoping language of the 2009 IBC (ICC 2009a):

> Detached one- and two-family dwellings and multiple single-family dwellings (townhouses) not more than three stories above grade plane in height with a separate means of egress and their accessory structures shall comply with the International Residential Code.

Therefore, primary guidance for regulatory requirements for the design and construction of buildings of interest in this Manual (e.g., one-and two-family detached dwellings) are based on the requirements specified in the IRC.

Generally, construction of residential buildings under the IRC need not involve a registered design professional, unless required by State law for the jurisdiction where the building is constructed. However, the building designer should be aware that engineered design is broadly permitted in the IRC and applicable even for a building structure with requirements contained entirely within the IRC, as stated in Section R301.1.3 (ICC 2009b):

> Engineered design in accordance with the International Building Code is permitted for all buildings and structures, and parts thereof, included in the scope of this code.

In certain cases and most coastal areas, however, the IRC requires structural elements to be "designed in accordance with accepted engineering practice." For example, engineered design of structural elements which fall outside the scope of requirements in the IRC such as building systems of excessive weight, elements of excessive length or height, or products not specifically addressed in the IRC is required. IRC Section R322.3.6 requires that construction documents be prepared and sealed by a registered design professional, and include documentation that the design and methods of construction to be used meet the applicable criteria of the IRC.

Buildings in regions of high wind, seismic, snow, and flood hazards as well as building elements outside of the range of limitations in the IRC require design beyond the IRC prescriptive provisions as follows:

- **Wind.** Buildings located where the basic wind speed equals or exceeds 110 miles per hour or where the IRC indicates special design for wind is required (wind speed triggers for the hurricane-prone region are based on mapped wind speeds in the 2012 IRC).

- **Seismic.** Buildings located in Seismic Design Category E.

- **Snow.** Buildings in regions with ground snow loads greater than 70 pounds per square foot.

- **Flood.** Buildings and structures constructed in whole or in part in coastal high hazard areas (including Zone V).

> **NOTE**
>
> The 2012 IRC replaces the previous basic wind speed map, Figure R301.2(4), with three new figures.
>
> - Figure R301.2(4)A presents a new map of basic wind speeds based on the ASCE 7-10 wind map data but converted to allowable-stress design (ASD) levels.
>
> - Figure R301.2(4)B provides shaded regions that indicate where wind speeds equal or exceed the scope of the IRC and use of recognized standards for wind design is required.
>
> - Figure R301.2(4)C indicates where the openings of buildings must be protected from wind-borne debris in accordance with ASTM E1996.
>
> Wind speed maps and triggers in the 2012 IRC are on an ASD basis, while wind speed maps and triggers in ASCE 7-10 are on a strength basis.

In addition to provisions of the IBC, applicable standards specifically recognized as accepted engineering practice for wind design within the IRC are: American Forest and Paper Association (AF&PA), *Wood Frame Construction Manual for One- and Two-Family Dwellings* (AF&PA 2012); ICC 600, *Standard for Residential Construction in High-Wind Regions* (ICC 2008a); ASCE 7-10, *Minimum Design Loads for Buildings and Other Structures* (ASCE 2010); and American Iron and Steel Institute (AISI), *Standard for Cold-Formed Steel Framing—Prescriptive Method For One- and Two-Family Dwellings with Supplement 2* (AISI 2007). For flood, ASCE 24-05, *Flood Resistant Design and Construction* (ASCE 2005), is specifically recognized within the IRC as an alternative to the flood design provisions of the IRC.

Engineered design requirements within both the IRC and IBC recognize ASCE 7 as the standard reference for minimum design loads due to hazards such as wind, flood, and seismic. As a result, within this Manual, provisions of ASCE 7 are used extensively for determination of minimum loads in accordance with engineered design requirements of the codes. For many portions of the Pacific, Great Lakes, and New England coasts, construction will generally fall within the prescriptive limits of the 2012 IRC and not require engineered design.

5.4 Best Practices for Exceeding Minimum NFIP Regulatory Requirements

This section presents best practices for exceeding NFIP minimum requirements. These best practices address the significant hazards present in Coastal A Zone and Zone V and are aimed at increasing the ability of coastal residential buildings to withstand natural hazard events. Refer to Section 5.2 for the minimum requirements of the NFIP regulations concerning buildings in Zone A and Zone V.

Table 5-2 in Section 5.4.3 summarizes the NFIP requirements and the best practices of this Manual regarding buildings in Zone A, Coastal A Zone, and Zone V.

5.4.1 Zone A

This Manual includes discussion of best practices for the design and construction of buildings in areas subject to coastal flooding, but focuses on Zone V and the Coastal A Zone (the portion of Zone A seaward of the LiMWA). However, development in the portion of Zone A landward of the LiMWA can benefit from many of the Zone V and Coastal A Zone design and construction practices included in this Manual. Designers seeking guidance regarding good practice for the design and construction of such buildings should consult local floodplain management, building, or code officials. Additional guidance can be found in FEMA 259, *Engineering Principles and Practices for Retrofitting Flood-Prone Residential Structures* (FEMA 2011); the IBC (ICC 2012a) and IRC (ICC 2012b); and the FEMA NFIP Technical Bulletins (available at http://www.fema.gov/plan/prevent/floodplain/techbul.shtm). This Manual recommends the provisions of ASCE 24 as best practices. These include, but are not limited to, the addition of freeboard in elevation requirements in Zone A (Figure 5-1).

5.4.2 Coastal A Zone and Zone V

As explained in Chapters 1 and 3 of this Manual, the NFIP regulations do not differentiate between the Coastal A Zone and the portion of Zone A that is landward of the LiMWA. Because Coastal A Zones may be subject to the types of hazards present in Zone V, such as wave effects, velocity flows, erosion, scour, and high winds, this Manual recommends that buildings in Coastal A Zones meet the NFIP regulatory requirements for Zone V buildings (i.e., the performance requirements concerning resistance to flotation, collapse, and lateral movement and the prescriptive requirements concerning elevation, foundation type, engineering certification of design and construction, enclosures below the lowest floor, and use of structural fill—see Section 5.2.3.3).

To provide a greater level of protection against the hazards in Coastal A Zone and Zone V, this Manual recommends the following as good practice for the siting, design, and construction of buildings in those zones:

- The building should be located landward of both the long-term erosion setback and the limit of base flood storm erosion, rather than simply landward of the reach of mean high tide.

- The bottom of the lowest horizontal structural member should be elevated above, rather than to, the BFE (i.e., provide freeboard—see Figure 5-2[b]).

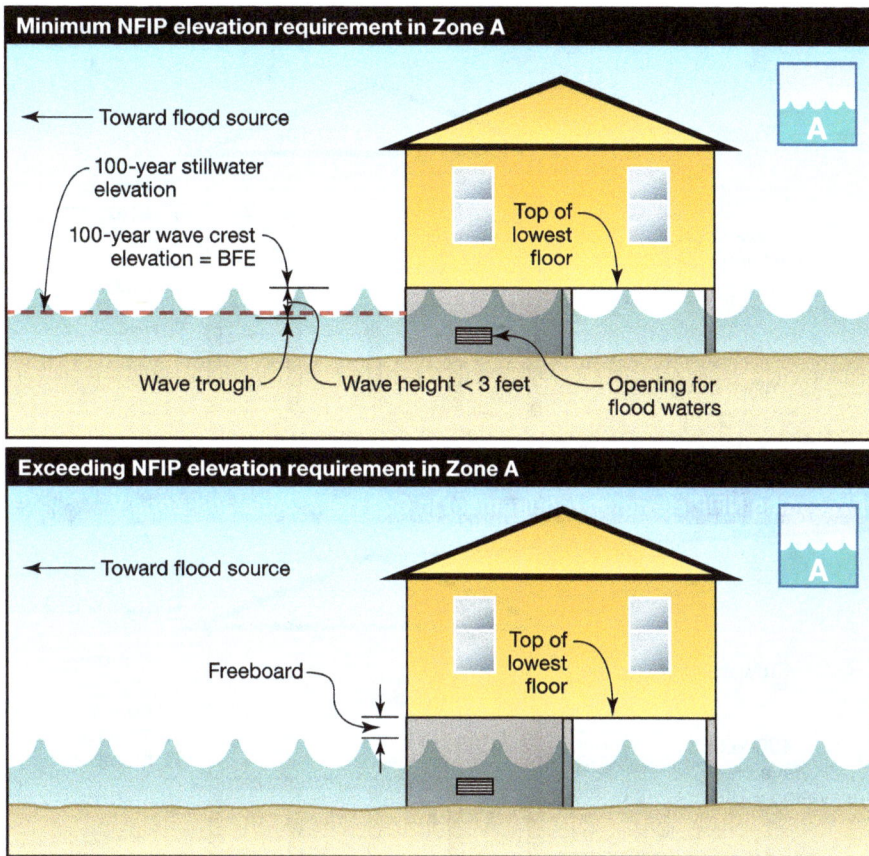

Figure 5-1.
Recommended elevation for buildings in Zone A compared to minimum requirements

- Open latticework, screening, or louvers should be used in lieu of breakaway walls in the space below the lowest floor, or, at a minimum, the use of solid breakaway walls should be minimized.

- In Zone V, the lowest horizontal structural member should be oriented perpendicular to the expected wave crest.

5.4.3   Summary

Table 5-2 summarizes NFIP regulatory requirements for Zone V, Coastal A Zone, and Zone A, and best practices for exceeding the requirements. These requirements and recommendations are in addition to the minimum building code requirements.

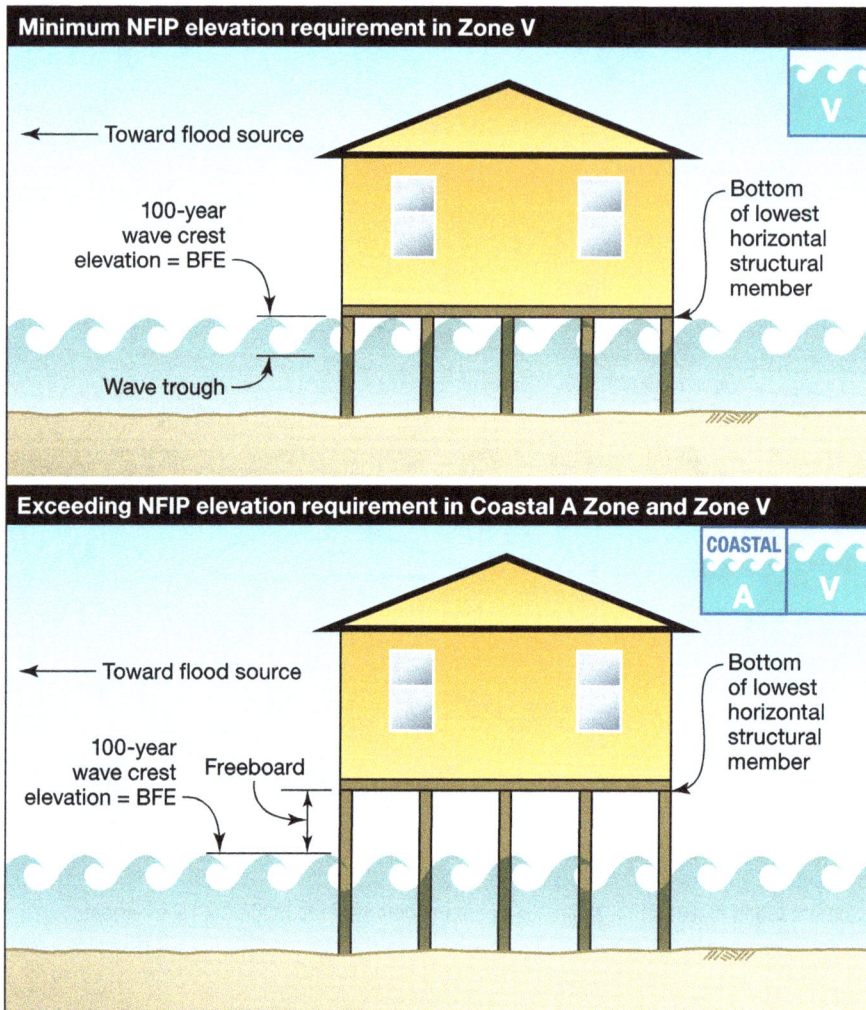

Minimum NFIP elevation requirement in Zone V

V

Toward flood source

100-year
wave crest
elevation = BFE

Wave trough

Bottom
of lowest
horizontal
structural
member

Exceeding NFIP elevation requirement in Coastal A Zone and Zone V

COASTAL
A   V

Toward flood source

100-year
wave crest
elevation = BFE

Freeboard

Bottom
of lowest
horizontal
structural
member

Figure 5-2.
Recommended elevation for buildings in Coastal A Zone and Zone V compared to minimum
requirements

Table 5-2. Summary of NFIP Regulatory Requirements and Recommendations for Exceeding the Requirements

	Zone V		Coastal A Zone		Zone A	
	Recommendations and Requirements[a]	Cross Reference[b]	Recommendations and Requirements	Cross Reference	Recommendations and Requirements	Cross Reference
GENERAL REQUIREMENTS						
Siting	**Recommendation:** Define and evaluate vulnerability to all coastal hazards, including short- and long-term erosion, and site building as far landward as possible. **Requirement:** New construction is landward of the reach of mean high tide. Manmade alterations of sand dunes and mangrove stands that increase potential flood damage are prohibited.	**NFIP:** 60.3(e)(3), 60.3(e)(7) **IRC:** R322.3.1 **IBC:** App. G401.2, App. G103.7 **ASCE 24:** 4.3 **FEMA P-55:** 2.3.2, Ch. 4, 7.5.1 **FEMA P-499:** 2.1, 2.2	**Recommendation:** Follow Zone V recommendations and requirements. **Requirement:** Buildings governed by IRC – meet Zone A requirements (unless authority having jurisdiction has adopted ASCE 24 for buildings governed by IRC). Buildings governed by IBC – follow Zone V requirements.	**IBC:** 1804.4 **ASCE 24:** 4.3 **FEMA P-55:** 2.3.2, Ch. 4 **FEMA P-499:** 2.1, 2.2	**Recommendation:** Site building outside of SFHA or on highest and most stable part of lot. **Requirement:** For floodways, fill is permitted only if it has been demonstrated that the fill will not result in any increase in flood levels during the base flood.	**NFIP:** 60.3(d)(3) **IRC:** R301.2.4, R322.1, R322.1.4.2 **IBC:** 1612.3.4, 1804.4, App. G 103.5, App. G 401.1 **ASCE 24:** 2.2 **FEMA P-55:** 2.3.2, Ch. 4
Design and Construction	**Recommendation:** Redundant and continuous load paths should be employed to transfer all loads to the ground. Designs should explicitly account for all design loads and conditions. **Requirement:** Building and foundation must be designed, constructed, and adequately anchored to prevent flotation, collapse, and lateral movement due to simultaneous wind and flood loads, including the effects of buoyancy.	**NFIP:** 60.3(a)(3)(i), 60.3(e)(4) **IRC:** R301.1, R301.2.4, R322.1, R322.3.3 **IBC:** 1603.1.7, 1604, 1605.2.2, 1605.3.1.2, 1612 **ASCE 7:** Ch. 5 **ASCE 24:** 1.5, Ch. 4 **FEMA P-55:** 2.3.3, 2.3.4, 5.4.2, Ch. 8, 9.1, 9.2 **FEMA P-499:** 3.1, 3.2, 3.3, 3.4, 4.1, 4.3 **Other:** FEMA P-550	**Recommendation:** Follow Zone V recommendations and requirements. **Requirement:** Building and foundation must be designed, constructed, and adequately anchored to prevent flotation, collapse, and lateral movement resulting from hydrodynamic and hydrostatic loads, including the effects of buoyancy.	**NFIP:** 60.3(a)(3)(i) **IRC:** R301.1, R301.2.4, R322.1.2, R322.2 **IBC:** 1603.1.7, 1604, 1605.2.2, 1605.3.1.2, 1612 **ASCE 7:** Ch. 5 **ASCE 24:** 1.5, Ch. 4 **FEMA P-55:** 2.3.3, 2.3.4, 5.4.2, Ch. 8, 9.1, 9.2 **FEMA P-499:** 3.1, 3.2, 3.3, 3.4, 3.5, 4.2, 4.2, 4.3 **Other:** FEMA P-550	**Recommendation:** Follow ASCE 24 requirements. **Requirement:** Building and foundation must be designed, constructed, and adequately anchored to prevent flotation, collapse, and lateral movement resulting from hydrodynamic and hydrostatic loads, including the effects of buoyancy.	**NFIP:** 60.3(a)(3)(i) **IRC:** R301.1, R301.2.4, R322.1.2, R322.2 **IBC:** 1603.1.7, 1604, 1605.2.2, 1605.3.1.2, 1612 **ASCE 7:** Ch. 5 **ASCE 24:** 1.5, 2.2 **FEMA P-55:** 2.3.3, 2.3.4, 5.4.1, Ch. 8, 9.1, 9.2 **Other:** FEMA P-550

Table 5-2. Summary of NFIP Regulatory Requirements and Recommendations for Exceeding the Requirements (continued)

	Zone V — Recommendations and Requirements[a]	Zone V — Cross Reference[b]	Coastal A Zone — Recommendations and Requirements	Coastal A Zone — Cross Reference	Zone A — Recommendations and Requirements	Zone A — Cross Reference
Flood Damage-Resistant Materials	**Recommendation:** Consider use of flood damage-resistant materials above BFE. **Requirement:** Structural and nonstructural building materials below the DFE must be flood damage-resistant.	**NFIP:** 60.3(a)(3)(ii) **IRC:** R322.1.8 **IBC:** 801.5, 1403.5 **ASCE 24:** Ch. 5 **FEMA P-55:** 5.2.3.1, 9.4 **FEMA P-499:** 1.7, 1.8, 4.3 **Other:** FEMA TB-2 and TB-8	**Recommendation:** Follow Zone V recommendations and requirements. **Requirement:** Structural and nonstructural building materials below the DFE must be flood damage-resistant.	**NFIP:** 60.3(a)(3)(ii) **IRC:** R322.1.8 **IBC:** 801.5, 1403.5 **ASCE 24:** Ch. 5 **FEMA P-55:** 5.2.3.1, 9.4 **FEMA P-499:** 1.7, 1.8, 4.3 **Other:** FEMA TB-2 and TB-8	**Recommendation:** Follow Zone V recommendations and requirements. **Requirement:** Structural and nonstructural building materials below the DFE must be flood damage-resistant.	**NFIP:** 60.3(a)(3)(ii) **IRC:** R322.1.8 **IBC:** 801.5,1403.5 **ASCE 24:** Ch. 5 **FEMA P-55:** 5.2.3.1, 9.4 **FEMA P-499:** 1.7, 1.8 **Other:** FEMA TB-2 and TB-8
Free of Obstructions	**Recommendation:** Use lattice, insect screening, or louvers instead of solid breakaway walls. **Requirement:** Open foundation required. The space below the lowest floor must be free of obstructions, or constructed with non-supporting breakaway walls, open lattice, or insect screening. Obstructions include any building element, equipment, or other fixed objects that can transfer flood loads to the foundation, or that can cause floodwaters or waves to be deflected into the building.	**NFIP:** 60.3(e)(5) **IRC:** R322.3.3 **IBC:** 1612.4 **ASCE 24:** 4.5.1 **FEMA P-55:** 5.2.3.3, 7.6.1.1.6, Table 7-3, Table 7-4, 10.5, 10.6 **FEMA P-499:** 1.2, 3.1, 8.1 **Other:** FEMA TB-5	**Recommendation:** Follow Zone V recommendation and requirement. **Requirement:** No limitations are imposed on obstructions below elevated floors unless the design is governed by IBC/ASCE 24 (in which case the lowest floor must be free of obstructions).	**IBC:** 1612.4 **ASCE 24:** 4.5.1 **FEMA P-55:** 5.2.3.3, 7.6.1.1.6, 10.5, 10.6 **FEMA P-499:** 1.2, 3.1, 8.1 **Other:** FEMA TB-5	**Recommendation:** If riverine flood velocities are high or large debris load is anticipated, open foundations are recommended. **Requirement:** None	**FEMA P-55:** 10.7, 10.9

Table 5-2. Summary of NFIP Regulatory Requirements and Recommendations for Exceeding the Requirements (continued)

	Zone V		Coastal A Zone		Zone A	
	Recommendations and Requirements[a]	Cross Reference[b]	Recommendations and Requirements	Cross Reference	Recommendations and Requirements	Cross Reference
ELEVATION						
Lowest Floor Elevation[e]	**Recommendation:** See Freeboard (additional height above required lowest floor elevation). **Requirement:** • NFIP: Bottom of the lowest horizontal structural member (LHSM)[d] of the lowest floor must be at or above the BFE. • IRC: Bottom of LHSM must be (a) at or above DFE if LHSM is parallel to direction of wave approach; or (b) at or above BFE plus 1 foot or DFE, whichever is higher, if LHSM is perpendicular to the direction of wave approach. • IBC/ASCE 24: Elevation based on orientation of LHSM and structure category. **Requirement:** See Lowest Floor Elevation	**NFIP:** 60.3(e)(4) **IRC:** R322.3.2, R332.1.5 **IBC:** 1612.4 **ASCE 24:** 1.5.2, 4.4 **FEMA P-55:** 5.2.3 **FEMA P-499:** 1.4	**Recommendation:** See Freeboard (additional height above required lowest floor elevation). **Requirement:** • NFIP: Top of floor must be at or above BFE. • IRC: Same as Zone A, plus 1 foot, if the LiMWA is delineated. • IBC/ASCE 24: Same as Zone V.	**NFIP:** 60.3(c)(2) **IRC:** R322.2.1, R322.1.5 **IBC:** 1612.4 **ASCE 24:** 1.5.2, 4.4 **FEMA P-55:** 5.2.3 **FEMA P-499:** 1.4	**Recommendation:** See Freeboard (additional height above required lowest floor elevation). **Requirement:** Top of floor must be at or above BFE.	**NFIP:** 60.3(c)(2) **IRC:** R322.2.1, R322.1.5 **IBC:** 1612.4 **ASCE 24:** 1.5.2, 2.3 **FEMA P-55:** 5.2.3 **FEMA P-499:** 1.4
Freeboard (additional height above required Lowest Floor Elevation)[e]	**Recommendation:** Elevate buildings higher than the required lowest floor elevation to provide more protection against flood damage and to reduce the cost of Federal flood insurance. **Requirement:** See Lowest Floor Elevation	**IRC:** R322.3.2 **IBC:** 1612.4 **ASCE 24:** 1.5.2, 4.4 **FEMA P-55:** 2.3.3, 5.4.2, 6.2.1, 7.5.2 (text box) **FEMA P-499:** 1.6 **Other:** NFIP Evaluation Study	**Recommendation:** Elevating building higher than the required lowest floor elevation provides more protection against flood damage and reduces the cost of Federal flood insurance. **Requirement:** See Lowest Floor Elevation	**IRC:** R322.2.1 **IBC:** 1612.4 **ASCE 24:** 1.5.2, 4.4 **FEMA P-55:** 2.3.3, 5.4.2, 6.2.1, 7.5.2 (text box) **FEMA P-499:** 1.6 **Other:** NFIP Evaluation Study	**Recommendation:** Elevating buildings higher than the required lowest floor elevation provides more protection against flood damage and reduces the cost of Federal flood insurance. **Requirement:** See Lowest Floor Elevation	**IBC:** 1612.4 **ASCE 24:** 1.5.2, 2.3 **FEMA P-55:** 2.3.3, 5.4.1, 6.2.1, 7.5.2 (text box) **FEMA P-499:** 1.6 **Other:** NFIP Evaluation Study

Table 5-2. Summary of NFIP Regulatory Requirements and Recommendations for Exceeding the Requirements (continued)

	Zone V		Coastal A Zone		Zone A	
	Recommendations and Requirements[a]	Cross Reference[b]	Recommendations and Requirements	Cross Reference	Recommendations and Requirements	Cross Reference
FOUNDATION						
Open Foundation	**Recommendation:** Follow requirement. **Requirement:** Open foundations (pilings or columns) are required.	**NFIP:** 60.3(e)(4) **IRC:** R322.3.3, R401.1 **IBC:** 1612.4 **ASCE 24:** 1.5.3, 4.5 **FEMA P-55:** 2.3.3, 5.2.3, 10.2, 10.3 **FEMA P-499:** 3.1, 3.2, 3.3, 3.4 **Other:** FEMA P-550	**Recommendation:** Follow Zone V requirement. **Requirement:**[e] Not required unless the design is governed by IBC/ASCE 24 (in which case an open foundation is required).	**IBC:** 1612.4 **ASCE 24:** 1.5.3, 4.5 **FEMA P-55:** 2.3.3, 5.2.3, 10.2, 10.3 **FEMA P-499:** 3.1, 3.2, 3.3, 3.4, 3.5 **Other:** FEMA P-550	**Recommendation:** If riverine flood velocities are high or large debris load is anticipated, open foundations are recommended. **Requirement:** None[e]	**IBC:** 1612.4 **ASCE 24:** 1.5.3, 2.4, 2.5 **FEMA P-55:** 2.3.3, 5.2.3, 10.2, 10.3 **FEMA P-499:** 3.5 **Other:** FEMA P-550
Solid Foundation Walls (including walls forming crawlspace, and stemwall foundations)	Not Permitted	**NFIP:** 60.3(e)(4) **IRC:** R322.3.3 **FEMA P-55:** 5.2.3.3, 7.6.1.1.6, 10.2, 10.3 **FEMA P-499:** 3.1, 3.5 **Other:** FEMA TB-5, FEMA P-550	**Recommendation:** Use open foundations. **Requirement:**[e] • NFIP: Solid foundation walls are required to have flood openings. • IRC: Wall height is limited, unless designed. • IBC/ASCE 24: Solid foundation walls are not permitted if design is governed by IBC/ASCE 24.	**NFIP:** 60.3(c)(5) **IRC:** R322.2.2, R322.2.3 **IBC:** 1612.4 **FEMA P-55:** 10.2, 10.3, 10.8 **FEMA P-499:** 3.1, 3.5 **Other:** FEMA P-550, FEMA TB-1	**Recommendation:** If velocities are high or debris load is anticipated, open foundations are recommended in lieu of elevation on solid walls. **Requirement:**[e] • NFIP: Solid foundation walls are required to have flood openings. • IRC: Wall height is limited, unless designed; walls are required to have flood openings. • IBC/ASCE 24: Solid foundation walls are required to have flood openings.	**NFIP:** 60.3(c)(5) **IRC:** R322.2.2, R322.2.3 **IBC:** 1612.4 **ASCE 24:** 2.6 **FEMA P-55:** 10.2, 10.3, 10.8 **FEMA P-499:** 3.1, 3.5 **Other:** FEMA P-550, FEMA TB-1

Table 5-2. Summary of NFIP Regulatory Requirements and Recommendations for Exceeding the Requirements (continued)

	Zone V		Coastal A Zone		Zone A	
	Recommendations and Requirements[a]	Cross Reference[b]	Recommendations and Requirements	Cross Reference	Recommendations and Requirements	Cross Reference
Structural Fill (including slab-on-grade foundation)	**Not Permitted**	**NFIP:** 60.3(e)(6) **IRC:** R322.3.2 **IBC:** 1612.4, 1804.4, App. G401.2 **ASCE 24:** 4.5.4 **FEMA P-55:** 5.2.3.3	**Recommendation:** Use open foundations. **Requirement:** If structural fill is used, compaction is necessary to meet requirements for stability during the base flood.	**NFIP:** 60.3(a)(3)(i) **IRC:** R322.1.2, R506 **IBC:** 1612.4, 1804.4 **ASCE 24:** 4.5.4 **FEMA P-55:** 10.3.1	**Recommendation:** If velocities are high or debris load is anticipated, open foundations are recommended in lieu of elevation on fill. **Requirement:** If structural fill is used, compaction is necessary to meet requirements for stability during the base flood.	**NFIP:** 60.3(a)(3)(i) **IRC:** R322.1.2, R506 **IBC:** 1612.4, 1804.4, App. G 401.1 **ASCE 24:** 2.4 **FEMA P-55:** 10.3.1
ENCLOSURES BELOW ELEVATED BUILDINGS						
Use of Enclosed Areas Below Elevated Lowest Floor[f]	**Recommendation:** Minimize use of enclosed areas to reduce damage to stored contents, and to reduce flood-borne debris. Avoid storage of damageable items and hazardous materials. **Requirement:** Enclosures are permitted only for parking of vehicles, building access, and storage.	**NFIP:** 60.3(e)(5) **IRC:** R322.3.5 **IBC:** 1612.4 **ASCE 24:** 4.6 **FEMA P-55:** 5.2.3.3 **FEMA P-499:** 8.1	**Recommendation:** Follow Zone V recommendations and requirements. **Requirement:** Enclosures are permitted only for parking of vehicles, building access, and storage.	**NFIP:** 60.3(c)(5) **IRC:** R322.2.2 **IBC:** 1612.4 **ASCE 24:** 4.6 **FEMA P-55:** 5.2.3.2 **FEMA P-499:** 8.1	**Recommendation:** Avoid storage of damageable items and hazardous materials in flood-prone spaces. **Requirement:** Enclosures are permitted only for parking of vehicles, building access, and storage.	**NFIP:** 60.3(c)(5) **IRC:** R322.2.2 **IBC:** 1612.4 **ASCE 24:** 2.6 **FEMA P-55:** 5.2.3.2
Walls of Enclosures[g]	**Recommendation:** Enclose areas with lattice, insect screening or louvers. Use flood openings to minimize collapse of solid breakaway walls under flood loads less than base flood loads.	**NFIP:** 60.3(e)(5) **IRC:** R322.3.4 **IBC:** 1612.4 **ASCE 24:** 4.6 **FEMA P-55:** 2.3.5, 5.2.3.2	**Recommendation:** Follow Zone V recommendations and requirements. **Requirement:** Solid foundation wall enclosures and solid breakaway wall enclosures must have flood openings.	**NFIP:** 60.3(c)(5) **IRC:** R322.2.2 **IBC:** 1612.4 **ASCE 24:** 4.6 **FEMA P-55:** 2.3.5, 5.2.3.2, 7.6.1.1.5	**Recommendation:** Follow requirement. **Requirement:** Walls of enclosures must have flood openings.	**NFIP:** 60.3(c)(5) **IRC:** R322.2.2 **IBC:** 1612.4 **ASCE 24:** 2.6 **FEMA P-55:** 2.3.5, 5.2.3.2, 7.6.1.1.5

Table 5-2. Summary of NFIP Regulatory Requirements and Recommendations for Exceeding the Requirements (continued)

	Zone V		Coastal A Zone		Zone A	
	Recommendations and Requirements[a]	Cross Reference[b]	Recommendations and Requirements	Cross Reference	Recommendations and Requirements	Cross Reference
Walls of Enclosures[g] (continued)	**Requirement:** Walls must be designed to collapse (break away) under flood loads to allow free passage of floodwaters without damaging the structure or supporting foundation system. Utilities and equipment must not be mounted on or pass through breakaway walls.	**FEMA P-499:** 8.1 **Other:** FEMA TB-9		**FEMA P-499:** 3.1, 3.5, 8.1 **Other:** FEMA TB-1 and TB-9		**FEMA P-499:** 3.1, 3.5 **Other:** FEMA TB-1
UTILITIES						
Electrical, Heating, Ventilation, Plumbing and Air Conditioning Equipment	**Recommendation:** Locate equipment on the landward side of building, and/or behind structural element. **Requirement:** Utilities and equipment must be located (elevated) and designed to prevent flood waters from entering and accumulating in components during flooding.	**NFIP:** 60.3(a)(3)(iv) **IRC:** R322.1.6, RM1301.1.1, RM1401.5, RM1601.4.9, RM1701.2, RM2001.4, RM2201.6, RG2404.7, RP2601.3, RP2602.2, RP2705.1, RP2101.5 **IBC:** 1403.5, 1403.6, 1612.4 **ASCE 24:** Ch. 7 **FEMA P-55:** Ch. 12 **FEMA P-499:** 8.3 **Other:** FEMA P-348, FEMA TB-5	**Recommendation:** Follow the Zone V recommendation and requirements. **Requirement:** Utilities and equipment must be located (elevated) and designed to prevent flood waters from entering and accumulating in components during flooding.	**NFIP:** 60.3(a)(3)(iv) **IRC:** R322.1.6, RM1301.1.1, RM1401.5, RM1601.4.9, RM1701.2, RM2001.4, RM2201.6, RG2404.7, RP2601.3, RP2602.2, RP2705.1, RP2101.5 **IBC:** 1403.5, 1612.4 **ASCE 24:** Ch. 7 **FEMA P-55:** Ch. 12 **FEMA P-499:** 8.3 **Other:** FEMA P-348, FEMA TB-5	**Recommendation:** Locate equipment on the landward or downstream side of building, and/ or behind structural element. **Requirement:** Utilities and equipment must be located (elevated) and designed to prevent flood waters from entering and accumulating in components during flooding.	**NFIP:** 60.3(a)(3)(iv) **IRC:** R322.1.6, RM1301.1.1, RM1401.5, RM1601.4.9, RM1701.2, RM2001.4, RM2201.6, RG2404.7, RP2601.3, RP2602.2, RP2705.1, RP2101.5 **IBC:** 1403.5, 1612.4 **ASCE 24:** Ch. 7 **FEMA P-55:** Ch. 12 **FEMA P-499:** 8.3 **Other:** FEMA P-348

Table 5-2. Summary of NFIP Regulatory Requirements and Recommendations for Exceeding the Requirements (continued)

	Zone V		Coastal A Zone		Zone A	
	Recommendations and Requirements[a]	Cross Reference[b]	Recommendations and Requirements	Cross Reference	Recommendations and Requirements	Cross Reference
Water Supply and Sanitary Sewerage Systems	**Recommendation:** Install shutoff valves to isolate water and sewer lines that extend into flood-prone areas. **Requirement:** Systems must be designed to minimize or eliminate infiltration of floodwaters into systems. Sanitary sewerage systems must be located to avoid impairment or contamination during flooding.	**NFIP:** 60.3(a)(5), 60.3(a)(6) **IRC:** R322.1.7, RP2602.2, RP3001.3 **IBC:** App. G401.3, App. G401.4, App. G701 **ASCE 24:** 7.3 **FEMA P-55:** Ch. 12 **FEMA P-499:** 8.3 **Other:** FEMA P-348, FEMA TB-5	**Recommendation:** Follow Zone V recommendation. **Requirement:** Systems must be designed to minimize or eliminate infiltration of floodwaters into systems. Sanitary sewerage systems must be located to avoid impairment or contamination during flooding.	**NFIP:** 60.3(a)(5), 60.3(a)(6) **IRC:** R322.1.7, RP2602.2, RP3001.3 **IBC:** App. G401.3, App. G401.4, App. G701 **ASCE 24:** 7.3 **FEMA P-55:** Ch. 12 **FEMA P-499:** 8.3 **Other:** FEMA P-348, FEMA TB-5	**Recommendation:** Follow requirement. **Requirement:** Systems must be designed to minimize or eliminate infiltration of floodwaters into systems. Sanitary sewerage systems must be located to avoid impairment or contamination during flooding.	**NFIP:** 60.3(a)(5), 60.3(a)(6) **IRC:** R322.1.7, RP2602.2, RP3001.3 **IBC:** App. G401.3, App. G401.4, App. G701 **ASCE 24:** 7.3 **FEMA P-55:** Ch. 12 **FEMA P-499:** 8.3 **Other:** FEMA P-348
CERTIFICATION						
Design Certifications (foundations, breakaway walls, flood openings)	**Recommendation:** Follow requirement. **Requirement:** Registered design professional must certify that the design and methods of construction are in accordance with accepted standards of practice for meeting design requirements, including design of breakaway walls if designed to fail under loads more than 20 pounds per square foot.	**NFIP:** 60.3(e)(4), 60.3(e)(5) **IRC:** R322.3.6 **IBC:** 1612.5(2.2) and (2.3) **FEMA P-55:** 5.2.2.3, 5.4.2 **FEMA P-499:** 1.5, 3.1, 8.1 **Other:** FEMA TB-9	**Recommendation:** Follow Zone V requirement.	**NFIP:** 60.3(c)(5) **IRC:** R322.2.2(2.2) **IBC:** 1612.5(1.2) **FEMA P-55:** 5.4.2 **FEMA P-499:** 1.5, 3.1, 8.1 **Other:** FEMA TB-1 and TB-9	**Recommendation:** Follow requirement. **Requirement:** Registered design professional must certify performance of engineered flood openings (flood openings that do not conform to prescriptive requirement).	**NFIP:** 60.3(c)(5) **IRC:** R322.2.2(2.2) **IBC:** 1612.5(1.2) **Other:** FEMA TB-1

Table 5-2. Summary of NFIP Regulatory Requirements and Recommendations for Exceeding the Requirements (continued)

	Zone V		Coastal A Zone		Zone A	
	Recommendations and Requirements[a]	Cross Reference[b]	Recommendations and Requirements	Cross Reference	Recommendations and Requirements	Cross Reference
Design Certifications (foundations, breakaway walls, flood openings) (continued)			**Requirement:** Registered design professional must certify performance of engineered flood openings (flood openings that do not conform to prescriptive requirement). If designs are governed by IBC or ASCE 24, registered design professional must certify that the design and methods of construction are in accordance with accepted standards of practice for meeting design requirements, including design of breakaway walls if designed to fail under loads more than 20 pounds per square foot.			
Certification of Elevation	**Recommendation:** Surveyed elevation of the bottom of the LHSM should be submitted when that member is placed and prior to further vertical construction, and re-surveyed and submitted prior to the final inspection. **Requirement:** Surveyed elevation of the bottom of the LHSM must be submitted to the community (as-built).	**NFIP:** 60.3(b)(5), 60.3(e)(2) **IRC:** R109.1.3, R322.1.10 **IBC:** 110.3.3, 1612.5(2.1) **FEMA P-499:** 1.4, 8.3 **Other:** NFIP FMB 467-1	**Recommendation:** Follow Zone V recommendations and requirements. **Requirement:** Surveyed elevation of the lowest floor must be submitted to the community (as-built).	**NFIP:** 60.3(b)(5) **IRC:** R109.1.3, R322.1.10 **IBC:** 110.3.3, 1612.5(1.1) **FEMA P-499:** 1.4, 8.3 **Other:** NFIP FMB 467-1	**Recommendation:** Surveyed elevation of the lowest floor should be submitted upon placement and prior to further vertical construction, and re-surveyed and submitted prior to the final inspection. **Requirement:** Surveyed elevation of the lowest floor must be submitted to the community (as-built).	**NFIP:** 60.3(b)(5) **IRC:** R109.1.3, R322.1.10 **IBC:** 110.3.3, 1612.5(1.1) **FEMA P-499:** 1.4, 8.3 **Other:** NFIP FMB 467-1

Table 5-2. Summary of NFIP Regulatory Requirements and Recommendations for Exceeding the Requirements (continued)

OTHER	Zone V Recommendations and Requirements[a]	Cross Reference[b]	Coastal A Zone Recommendations and Requirements	Cross Reference	Zone A Recommendations and Requirements	Cross Reference
Non-Structural Fill	**Recommendation:** Minimize use of non-structural fill if flow diversion, wave runup, or reflection are concerns. Non-structural fill should be similar to existing soils where possible. **Requirement:** Minor quantities can be used for site grading, landscaping and drainage, and to support parking slabs, patios, walkways and pool decks. Non-structural fill can be used for dune construction or reconstruction. Non-structural fill must not prevent the free passage of floodwater and waves beneath elevated buildings, or lead to building damage through flow diversion or wave runup or reflection.	**NFIP:** 60.3(e)(5) **IRC:**, R322.3.2 **ASCE 24:** 4.5.4 **FEMA P-55:** 5.2.3.3 **Other:** FEMA TB-5	**Recommendation:**[h] Follow Zone V recommendation. **Requirement:** None	**IRC:** R322.3.2 **ASCE 24:** 4.5.4 **Other:** FEMA TB-5	**Recommendation:** Follow requirement. **Requirement:**[h,i] Encroachments into floodways are permitted only if it is demonstrated that the encroachment will not result in any increase in flood levels during the base flood.	**NFIP:** 60.3(d)(3) **IRC:** R301.2.4, R322.1, R322.1.4.2 **IBC:** 1612.3.4, 1804.4, App. G 103.5, App. G 401.1 **ASCE 24:** 2.2

Table 5-2. Summary of NFIP Regulatory Requirements and Recommendations for Exceeding the Requirements (continued)

	Zone V		Coastal A Zone		Zone A	
	Recommendations and Requirements(a)	Cross Reference(b)	Recommendations and Requirements	Cross Reference	Recommendations and Requirements	Cross Reference
Decks, Concrete Pads, Patios	**Recommendation:** Decks should be built using the same foundation as the main building, or cantilevered from the main building. Decks, pads, and patios should be designed to minimize the creation of large debris in the event of failure. **Requirement:** If structurally attached to buildings, decks, concrete pads and patios must be elevated.	**NFIP:** 60.3(e)(3) **IRC:** R322.3.3 **ASCE 24:** 4.8, 9.2 **FEMA P-55:** 9.5 **FEMA P-499:** 8.2 **Other:** FEMA TB-5	**Recommendation:** Follow Zone V recommendations. **Requirement:** If located below the DFE, decks, concrete pads, patios and similar appurtenances must be stable under flood loads.	**NFIP:** 60.3(a)(3) **ASCE 24:** 4.8, 9.2 **FEMA P-499:** 8.2 **Other:** FEMA TB-5	**Recommendation:** Follow requirement. **Requirement:** If located below the DFE, decks, concrete pads, patios and similar appurtenances must be stable under flood loads.	**NFIP:** 60.3(a)(3) **ASCE 24:** 9.2 **FEMA P-499:** 8.2
Swimming Pools	**Recommendation:** Pool should be located as far landward as possible and should be oriented in such a way that flood forces are minimized. **Requirement:** Swimming pools and pool decks must be stable under flood loads and elevated, designed to break away during the design flood or be sited to remain in-ground without obstructing flow that results in damage to adjacent structures.	**NFIP:** 60.3(e)(3) **IRC:** R322.3.3, App. G101.2 **ASCE 24:** 9.5 **FEMA P-55:** 9.5 **FEMA P-499:** 8.2 **Other:** FEMA TB-5	**Recommendation:** Follow Zone V recommendation. **Requirement:** Swimming pools and pool decks must be stable under flood loads.	**NFIP:** 60.3(a)(3) **IRC:** App. G101.2 **ASCE 24:** 9.5 **FEMA P-499:** 8.2	**Recommendation:** Follow requirement. **Requirement:** Swimming pools and pool decks must be stable under flood loads.	**NFIP:** 60.3(a)(3) **IRC:** App. G101.2 **ASCE 24:** 9.5

Table 5-2. Summary of NFIP Regulatory Requirements and Recommendations for Exceeding the Requirements (concluded)

	Zone V Recommendations and Requirements[a]	Zone V Cross Reference[b]	Coastal A Zone Recommendations and Requirements	Coastal A Zone Cross Reference	Zone A Recommendations and Requirements	Zone A Cross Reference
Tanks Associated with Building Utilities	**Recommendation:** Locate above-ground tanks on the landward side of buildings and raise inlets, fill openings, and vents above the DFE. Install underground tanks below the eroded ground elevation. **Requirement:** Above-ground tanks must be elevated.	**NFIP:** 60.3(e)(3) **IRC:** R2201.6 **IBC:** App. G701 **ASCE 24:** 7.4.1 **FEMA P-499:** 8.3 **Other:** FEMA TB-5, FEMA P-348	**Recommendation:** Follow Zone V recommendations. **Requirement:** Tanks must be elevated or anchored to be stable under flood loads, whether above-ground or underground.	**NFIP:** 60.3(a)(3) **IRC:** R2201.6 **IBC:** App. G701 **ASCE 24:** 7.4.1 **FEMA P-499:** 8.3 **Other:** FEMA TB-5, FEMA P-348	**Recommendation:** Locate above-ground tanks on the landward or downstream side of buildings and raise inlets, fill openings, and vents above the DFE. **Requirement:** Tanks must be elevated or anchored to be stable under flood loads, whether above-ground or underground.	**NFIP:** 60.3(a)(3) **IRC:** R2201.6 **IBC:** App. G701 **ASCE 24:** 7.4.1 **Other:** FEMA P-348
Sustainable Design	**Recommendation:** Building for natural hazards resistance reduces the need to rebuild and is a sustainable design approach. Verify that other green building practices do not reduce the building's ability to resist flood loads or other natural hazards. **Requirement:** Meet overall NFIP performance requirements.	**FEMA P-55:** 7.7 **Other:** FEMA P-798, ICC 700	**Recommendation:** Follow Zone V recommendation. **Requirement:** Meet overall NFIP performance requirements.	**FEMA P-55:** 7.7 **Other:** FEMA P-798, ICC 700	**Recommendation:** Follow Zone V recommendation. **Requirement:** Meet overall NFIP performance requirements.	**FEMA P-55:** 7.7 **Other:** FEMA P-798, ICC 700

Table 5-2 Notes:

(a) Individual States and communities may enforce more stringent requirements that supersede those summarized here. ***Exceeding minimum NFIP requirements will provide increased flood protection and may result in lower flood insurance premiums.***

(b) The references in this section cite the latest available publications at the time of publication of this Manual. The specific editions of these references are:

- ASCE 7: ASCE 7-10, *Minimum Design Loads for Buildings and Other Structures*
- ASCE 24: ASCE 24-05, *Flood Resistant Design and Construction*
- IBC: *2012 International Building Code.* Appendix G includes provisions for flood-resistant construction. The provisions in IBC Appendix G are not mandatory unless specifically referenced in the adopting ordinance. Many States have not adopted Appendix G. Section references are the same as 2009 IBC.
- ICC 700: *National Green Building Standard* (ICC 2008b)
- IRC: *2012 International Residential Code for One- and Two-Family Dwellings.* Section references are the same as 2009 IRC.
- FEMA P-55: Specific sections or chapters of this Manual; FEMA P-55, *Coastal Construction Manual* (2011)
- FEMA P-348: 1999 Edition of FEMA P-348, *Protecting Building Utilities from Flood Damage*
- FEMA P-499: Specific fact sheets in the 2010 edition of FEMA P-499, *Home Builder's Guide to Coastal Construction Technical Fact Sheet Series*
- FEMA P-550: FEMA P-550, *Recommended Residential Construction for Coastal Areas* (Second Edition, 2009)
- FEMA P-798: *Natural Hazards and Sustainability for Residential Buildings* (2010)
- FEMA TB: Specific numbered FEMA NFIP Technical Bulletins (available at http://www.fema.gov/plan/prevent/floodplain/techbul.shtm)
- NFIP: *U.S. Code of Federal Regulations* – 44 CFR § 60.3 "Flood plain management criteria for flood-prone areas." Current as of June 30, 2011.
- NFIP Evaluation Study: *Evaluation of the National Flood Insurance Program's Building Standards* (American Institutes for Research 2006)
- NFIP FMB 467-1: *Floodplain Management Bulletin on the NFIP Elevation Certificate.* Note that this bulletin was published in 2004, while the Elevation Certificate (FEMA Form 81-31) has been updated since 2004, and is updated periodically.

(c) State or community may regulate to a higher elevation (DFE).

(d) LHSM = Lowest horizontal structural member.

(e) Some coastal communities require open foundations in Zone A.

(f) There are some differences between what is permitted under floodplain management regulations and what is covered by NFIP flood insurance. Building designers should be guided by floodplain management requirements, not by flood insurance policy provisions.

(g) Some coastal communities prohibit breakaway walls and allow only open lattice or screening.

(h) Placement of nonstructural fill adjacent to buildings in Zone AO in coastal areas is not recommended.

(i) Some communities may allow encroachments to cause a 1-foot rise in the flood elevation, while others may allow no rise.

5.5   References

AF&PA (American Forest & Paper Association). 2012. *Wood Frame Construction Manual for One- and Two-Family Dwellings.* WFCM-12.

AISI (American Iron and Steel Institute). 2007. *Standard for Cold-Formed Steel Framing - Prescriptive Method for One- and Two-Family Dwellings with Supplement.* (AISI S230-07w/S2-08).

American Institutes for Research. 2006. *Evaluation of the National Flood Insurance Program's Building Standards.* October.

APA (American Planning Association). 2002. *Growing Smart Legislative Guidebook, Model Statutes for Planning and the Management of Change.* January.

ASCE (American Society of Civil Engineers). 2005. *Flood Resistant Design and Construction.* ASCE Standard ASCE 24-05.

ASCE. 2010. *Minimum Design Loads for Buildings and Other Structures.* ASCE Standard ASCE 7-10.

ASTM (ASTM International). 1996. *Standard Specification for Performance of Exterior Windows, Curtain Walls, Doors, and Impact Protective Systems Impacted by Windborne Debris in Hurricanes.* ASTM E1996.

FEMA (Federal Emergency Management Agency). 1991. *Answers to Questions About Substantially Damaged Buildings.* National Flood Insurance Program Community Assistance Series. FEMA 213. May.

FEMA. 1996. *Corrosion Protection for Metal Connectors in Coastal Areas for Structures Located in Special Flood Hazard Areas in accordance with the National Flood Insurance Program.* Technical Bulletin 8-96.

FEMA. 1999. *Protecting Building Utilities from Flood Damage.* FEMA P-348.

FEMA. 2004. *Floodplain Management Bulletin on the Elevation Certificate.* NFIP Bulletin 467-1. May.

FEMA. 2008a. *Design and Construction Guidance for Breakaway Walls Below Elevated Buildings Located in Coastal High Hazard Areas.* NFIP Technical Bulletin 9.

FEMA. 2008b. *Flood Damage-Resistant Materials Requirements for Buildings Located in Special Flood Hazard Areas.* NFIP Technical Bulletin 2.

FEMA. 2008c. *Free-of-Obstruction Requirements for Buildings Located in Coastal High Hazard Areas.* NFIP Technical Bulletin 5.

FEMA. 2008d. *Openings in Foundation Walls and Walls of Enclosures Below Elevated Buildings in Special Flood Hazard Areas.* NFIP Technical Bulletin 1.

FEMA. 2009a. *Protecting Manufactured Homes from Floods and Other Hazards, A Multi-Hazard Foundation and Installation Guide.* FEMA P-85. November.

FEMA. 2009b. *Recommended Residential Construction for Coastal Areas: Building on Strong and Safe Foundations.* FEMA P-550. December.

FEMA. 2010a. *Substantial Improvement/Substantial Damage Desk Reference.* FEMA P-758. May.

FEMA. 2010b. *The FEMA Substantial Damage Estimator (SDE).* FEMA P-784. June.

FEMA. 2010c. *Home Builder's Guide to Coastal Construction Technical Fact Sheet Series.* FEMA P-499. December.

FEMA. 2010d. *Natural Hazards and Sustainability for Residential Buildings.* FEMA P-798. September.

FEMA. 2011. *Engineering Principles and Practices for Retrofitting Flood-Prone Residential Structures.* FEMA P-259, Third Edition.

ICC (International Code Council). 2008a. *Standard for Residential Construction in High-Wind Regions, ICC 600.* Birmingham, AL.

ICC. 2008b. *National Green Building Standard, ICC 700.* Birmingham, AL.

ICC. 2009a. *International Building Code.* Birmingham, AL.

ICC. 2009b. *International Residential Code for One- and Two-Family Dwellings.* Birmingham, AL.

ICC. 2012a. *International Building Code.* Birmingham, AL.

ICC. 2012b. *International Residential Code for One- and Two-Family Dwellings.* Birmingham, AL.

ICC. 2012c. *International Existing Building Code.* Birmingham, AL.

ICC. 2012d. *International Mechanical Code.* Birmingham, AL.

ICC. 2012e. *International Plumbing Code.* Birmingham, AL.

ICC. 2012f. *International Private Sewage Disposal Code.* Birmingham, AL.

ICC. 2012g. *International Fuel Gas Code.* Birmingham, AL.

ICC. 2012h. *International Fire Code.* Birmingham, AL.

Morris, M. 1997. *Subdivision Design in Flood Hazard Areas.* Planning Advisory Service Report Number 473. American Planning Association.

NFPA (National Fire Protection Association). 2012. *Building Construction and Safety Code.* NFPA 5000.

Schwab, J.; K C. Topping; C. C. Eadie; R. E. Deyle; R. A. Smith. 1998. *Planning for Post-Disaster Recovery and Reconstruction.* Planning Advisory Service Report Number 483/484. American Planning Association.

Schwab, J. 2010. *Hazard Mitigation: Integrating Best Practices into Planning.* APA Planning Advisory Service Report Number 560.

Tung, C. C.; B. Kasal; S. M. Rogers, Jr.; S. C. Yeh. 1999. *Behavior of Breakaway Walls Subjected to Wave Forces: Analytical and Experimental Studies.* UNC-SG-99-03. North Carolina Sea Grant, North Carolina State University. Raleigh, NC.

Code of Federal Regulations "Flood plain management criteria for flood-prone areas." 44 CFR § 60.3. June 30, 2011.

Fundamentals of Risk Analysis and Risk Reduction

A successful building design incorporates elements of risk assessment, risk reduction, and risk management. Building success as defined in Chapter 1 can be met through various methods, but they all have one thing in common: careful consideration of natural hazards and use of siting, design, construction, and maintenance practices to reduce damage to the building. Designing in areas subject to coastal hazards requires an increased standard of care. Designers must also be knowledgeable about loading requirements in coastal hazard areas and appropriate ways to handle those loads. Failure to address even one of these concerns can lead to building damage, destruction, or loss of use. Designers should remember that the lack of building damage during a high-probability (low-intensity) wind, flood, or other event cannot be construed as a building success—success can only be measured against a design event or a series of lesser events with the cumulative effect of a design event.

> **CROSS REFERENCE**
>
> For resources that augment the guidance and other information in this Manual, see the Residential Coastal Construction Web site (http://www.fema.gov/rebuild/mat/fema55.shtm).

A critical component of successful building construction in coastal environments is accurately assessing the *risk* from natural hazards and then reducing that risk as much as possible. Accurate risk assessment and risk reduction are directly tied to correctly identifying natural hazards relevant to the building site. Before beginning the design process, it is important to understand and identify the natural hazard risks associated with a particular site, determine the desired level of protection from those hazards, and determine how best to manage *residual risk*. Design professionals must communicate these concepts to building owners so they can determine if the level of residual risk is acceptable or whether it would be cost-beneficial to further increase the hazard resistance of the building, and thereby reduce the residual risk. Once the desired level of protection and the residual risk have been evaluated by the designer and the owner, the information in Volume II can be used to incorporate appropriate forces and loads into a successful hazard-resistant design.

> **TERMINOLOGY**
>
> **RISK:**
> Potential losses associated with a hazard, defined in terms of expected probability and frequency, exposure, and consequences.
>
> **RESIDUAL RISK:**
> The level of risk that is not offset by hazard-resistant design or insurance, and that must be accepted by the property owner.

6.1 Assessing Risk

A hazard-resistant building design begins with a proper risk assessment. Building success can only be achieved by successfully identifying and managing natural hazard risks. Designing a successful building requires an understanding of the magnitude of the hazards and how frequently the building may be subjected to these hazards. This information is used to assess the potential exposure of the building to these hazards, i.e., the risk to the building. For the purposes of this Manual, *risk assessment* is the process of quantifying the total risk to a coastal building from all significant natural hazards that may impact the building.

Designers should be well informed with current hazard and risk information and understand how risk affects their design decisions and the requirements of the client. Designers should:

- Obtain the most up-to-date published hazard data to assess the vulnerability of a site, following the steps outlined in Section 4.3.

- Conduct or update a detailed risk assessment if there is reason to believe that physical site conditions have changed significantly since the hazard data were published or published hazard data is not representative of a site.

- Review or revise an existing risk assessment if there is reason to believe that physical site conditions will change significantly over the expected life of a structure or development of the site (see Section 3.7).

- After a risk assessment is completed, the designer should review siting and design options that will mitigate the effects of the identified hazards. The building owner may not find the amount of damage or loss of function acceptable, and the designer should work with the building owner to mitigate the risk to an acceptable level.

6.1.1 Identifying Hazards for Design Criteria

Coastal areas are subject to many hazards, including distinct events such as hurricanes, coastal storms, earthquakes, and earthquake-induced landslides and tsunamis. Coastal hazards also include continuous, less obvious coastal phenomena, such as long-term erosion, shoreline migration, and the corrosion and decay of building materials. The effects of hazards associated with distinct events are often immediate, severe,

> **CROSS REFERENCE**
>
> Chapter 7 presents an introduction to Volume II and a summary of the insurance and financial implications of design decisions.

and easily visible, while those associated with slow-onset, long-term processes are more likely to become apparent only over time. Manmade structures such as bulkheads, dams, dikes, groins, jetties, levees, and seawalls may also be present in coastal areas and the effects of these structures on nearby buildings must be considered.

CROSS REFERENCE

For information on identifying coastal hazards, refer to Chapter 3.

For siting considerations, refer to Chapter 4.

For discussion of codes and standards, refer to Chapter 5.

The designer must determine which specific hazards will affect a particular site and the vulnerability of the site to the identified natural hazards. Not all sites have the same hurricane exposure, erosion exposure, or seismic risk. The exposure of the building to these natural hazards should be evaluated and incorporated into the design criteria. The designer must first focus on code compliance. By following code provisions and NFIP regulations for flood, wind, and seismic design, the immediately understood and quantified hazards are mitigated to a certain degree. To fully understand the risk at a particular site, the designer should then study the risk associated with an above-design-level event. Finally, the designer should consider mitigation solutions to long-term issues such as erosion, subsidence, and sea level rise.

The designer should also address the possibility of unlikely events such as a levee failure (when appropriate). While such events may seem very unlikely or improbable to the owner, it is important that designers review flood maps, flood studies, and historical events to understand the risks to the building and how to best manage them.

Additionally, cumulative effects of multiple hazards should be considered. For example, hurricane-induced wind and flooding impacts may be exacerbated by sea level rise or subsidence. Designing buildings to resist these forces may present numerous challenges and therefore, these issues should be carefully evaluated.

6.1.2 Probability of Hazard Occurrence and Potential Consequences

Understanding the probabilities and the consequences of building damage or failure will help designers determine the level of natural hazard resistance they seek in the building design and better quantify the risk. Flood, wind, and seismic events have been studied and modeled with varying degrees of accuracy for centuries. Careful study of each of these hazards has resulted in a notable historical record of both the frequency and intensity of those events. The historic frequency of events with different intensities allows mathematical analysis of the events and the development of probabilities of future events. The probability of future events occurring can be used to predict the potential consequences of building design choices.

For instance, understanding the probability that a site will experience a specific wind speed allows a designer to carefully design the building for that wind speed and understand the wind risk to that building. The designer can also consult with the owner on the level of wind protection incorporated into the building design and help them determine how to manage the residual risk. Residual risk will be present because storm events that result in greater-than-design wind speeds can occur. Based on the owner's level of acceptance to risk, the owner may then decide to seek a higher level of building performance or purchase insurance to reduce the residual risk.

Designers must determine the probability of occurrence of each type of hazard event over the life of the structure and evaluate how often it might occur. The frequency of the occurrence of a natural hazard is referred to in most design codes and standards as the ***recurrence interval***. The probability of the occurrence of severe events should be evaluated over the life of the structure, and the consequences of their occurrence should be addressed in the design. While more frequent and less severe events may not have the same drastic consequences as less frequent but more severe events, they should still be identified and assessed in the risk assessment. In contrast, some events may be so severe and infrequent that it is likely not cost-effective to design the building to withstand them.

In most coastal areas of the United States, buildings must meet minimum regulatory and code requirements intended to provide protection from natural hazard events of specified magnitudes. These events are usually identified according to their recurrence intervals. For instance, the base flood used by the NFIP is associated with a recurrence interval of 100 years, the basic wind speed for Risk Category II structures in ASCE 7-10 is associated with a recurrence interval of 700 years, and the return interval for earthquake design is 2,500 years.

After identifying the recurrence interval of a natural hazard event or design event (through codes, standards, or other design criteria) the designer can determine the probability of one or more occurrences of that event or a larger event during a specified period, such as the expected lifespan of the building.

Table 6-1 illustrates the probability of occurrence for natural hazard events with recurrence intervals of 10, 25, 50, 100, 500, and 700 years. Of particular interest in this example is the event with a 100-year recurrence interval because it serves as the basis for the floodplain management and insurance requirements of the NFIP regulations, and floodplain regulations enforced by local governments. The event with a 100-year recurrence interval has a 1 percent probability of being equaled or exceeded over the course of 1 year (referred to as the 1-percent-annual-chance flood event). As the period increases, so does the probability that an event of this magnitude or greater will occur. For example, if a house is built to the 1-percent-annual-chance flood level (often referred to as the 100-year flood level), the house has a 26 percent chance of being flooded during a 30-year period, equivalent to the length of a standard mortgage (refer to the bolded cells in Table 6-1). Over a 70-year period, which may be assumed to be the useful life of many buildings, the home has a 51 percent chance of being flooded (refer to the bolded cells in Table 6-1). The same principle applies to other natural hazard events with other recurrence intervals.

> **WARNING**
>
> Designers of structures along Great Lakes shorelines, if they are using Table 6-1 to evaluate flood probabilities, should be aware that the table may underestimate actual probabilities during periods of high lake levels. For example, Potter (1992) calculated that during rising lake levels in 1985, Lake Erie had a 10 percent probability of experiencing a 100-year flood event in the next 12 months (versus 1 percent as shown in Table 6-1).

Table 6-1. Probability of Natural Hazard Event Occurrence for Various Periods of Time

Length of Period (Years)	Frequency – Recurrence Interval					
	10-Year	25-Year	50-Year	100-Year	500-Year	700-Year
1	10%	4%	2%	1%	0.2%	0.1%
10	65%	34%	18%	10%	2%	1%
20	88%	56%	33%	18%	4%	3%
25	93%	64%	40%	22%	5%	4%
30	96%	71%	45%	26%	6%	4%
50	99+%	87%	64%	39%	10%	7%
70	99.94+%	94%	76%	51%	13%	10%
100	99.99+%	98%	87%	63%	18%	13%

The percentages shown represent the probabilities of one or more occurrences of an event of a given magnitude or larger within the specified period. The formula for determining these probabilities is $P_n = 1-(1-P_a)^n$, where P_a = the annual probability and n = the length of the period.

The bold blue text in the table reflects the numbers used in the example in this section.

6.2   Reducing Risk

Once the risk has been assessed, the next step is to decide how to best mitigate the identified hazards. The probability of a hazard event occurrence is used to evaluate risk reduction strategies and determine the level of performance to incorporate into the design. The chance of severe flooding, high-wind events, or a severe earthquake can dramatically affect the design methodology, placement of the building on the site, and materials selected. Additionally, the risk assessment and risk reduction strategy must account for the short- and long-term effects of each hazard, including the potential for cumulative effects and the combination of effects from different hazards. Overlooking a hazard or underestimating its long-term effects can have disastrous consequences for the building and its owner.

WARNING

Meeting minimum regulatory and code requirements for the siting, design, and construction of a building does not guarantee that the building will be safe from all hazard effects. Risk to the building still exists. It is up to the designer and building owner to determine the amount of acceptable risk to the building.

Although designers have no control over the hazard forces, the siting, design, construction, and maintenance of the building are largely within the control of the designer and owner. The consequences of inadequately addressing these design items are the impetus behind the development of this Manual. *Risk reduction* is comprised of two aspects: *physical risk reduction* and *risk management through insurance*.

Eliminating all risk is impossible. Risk reduction, therefore, also includes determining the *acceptable level of residual risk*. Managing risk, including identifying acceptable levels of residual risk, underlies the entire coastal construction process. The initial, unmitigated risk is reduced through a combination of floodplain ordinances, building codes, best practices construction, and insurance. Each risk reduction element decreases the residual risk; the more elements that are applied, the smaller the remaining residual risk. Figure 6-1 shows the general level of risk reduction after each risk reduction element is applied.

Figure 6-1.
Initial risk is reduced to residual risk through physical and financial risk reduction elements

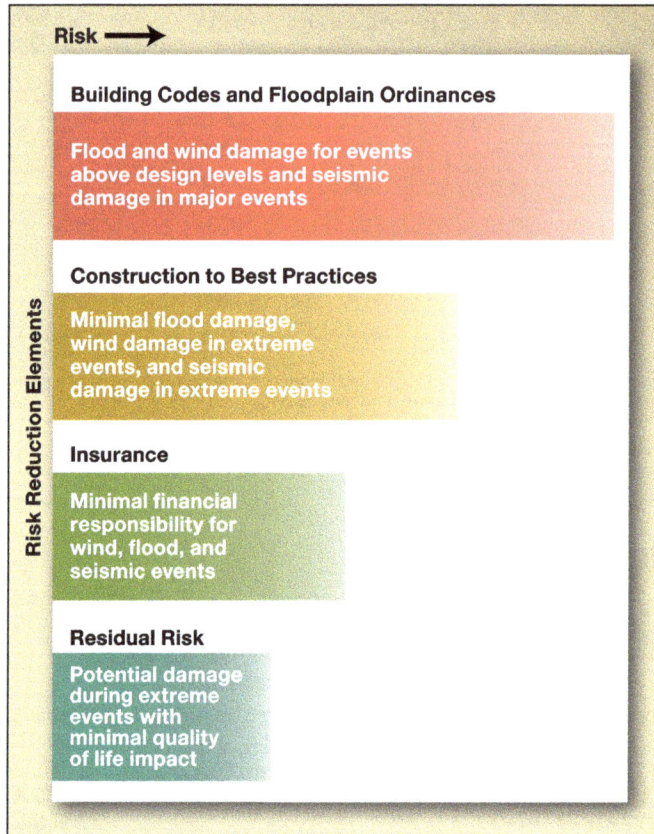

Risk ⟶

Building Codes and Floodplain Ordinances

Flood and wind damage for events above design levels and seismic damage in major events

Construction to Best Practices

Minimal flood damage, wind damage in extreme events, and seismic damage in extreme events

Insurance

Minimal financial responsibility for wind, flood, and seismic events

Residual Risk

Potential damage during extreme events with minimal quality of life impact

Risk Reduction Elements

6.2.1 Reducing Risk through Design and Construction

Building codes and Federal, State, and local regulations establish minimum requirements for siting, design, and construction. Among these are requirements that buildings be constructed to withstand the effects of natural hazards with specified recurrence intervals (e.g., 100-year for flood, 700-year for wind, 2,500-year for earthquake). Therefore, when building codes and regulatory requirements are met, they can help reduce the vulnerability of a building to natural hazards

CROSS REFERENCE

Chapter 5 presents information on building codes and standards for coastal construction.

and, in a sense, provide a baseline level of risk reduction. However, *meeting minimum regulatory and code requirements leaves a certain level of residual risk* that can and should be reduced through design and construction of the best practices described in this Manual.

Design decisions including elevation, placement and orientation of the building on the site, size and shape of the building, and the materials and methods used in its construction all affect a building's vulnerability to natural hazard events. However, these decisions can also affect initial and long-term costs (see Section 7.5), aesthetic qualities (e.g., the appearance of the finished building, views from within), and convenience for the homeowner (e.g., accessibility). The tradeoffs among these factors involve objective and subjective considerations that are often difficult to quantify and likely to be assessed differently by developers, builders, homeowners, and community officials. The cost of siting and design decisions must be balanced with the amount of protection from natural hazards provided.

6.2.1.1   Factors of Safety and Designing for Events that Exceed Design Minimums

Codes and standards require minimum levels of protection from natural hazards, including a minimum factor of safety. Factors of safety are designed to account for unknowns in the prediction of natural hazards and variability in the construction process and construction materials. Since the designer may have limited control over these factors it is important that they not only embrace the minimum factors of safety, but determine whether a *higher factor of safety* should be incorporated into the design to improve the hazard resistance of buildings. Such decisions can often result in other benefits besides increased risk reduction such as potential reduced insurance premiums and improved energy efficiency (see Chapter 7). The designer should also evaluate what the consequences would be to the building if the minimum design conditions were exceeded by a natural hazard event.

When beginning the design process, it is important to determine the building's *risk category* as defined in ASCE 7-10 and the 2012 IBC. A building's risk category is based on the risk to human life, health, and welfare associated with potential damage or failure of the building. The factors of safety incorporated into the design criteria increase as the risk category increases. These risk categories dictate which design event is used when calculating performance expectations of the building, specifically the loads the building is expected to resist. The risk categories from ASCE 7-10 are summarized as:

> **NOTE**
>
> ASCE 7-10 and the 2012 IBC introduced the term *risk categories*. Risk categories are called "occupancy categories" in previous editions. The broad categories in ASCE 7-10 are intended to represent the specific listings in the 2012 IBC. The descriptions provided in this Manual are broad, and both ASCE 7-10 and the 2012 IBC should be consulted to determine risk category.

- **Category I.** Buildings and structures that are normally unoccupied, such as barns and storage sheds, and would likely result in minimal risk to the public in the event of failure.

- **Category II.** All buildings and structures that are not classified by the other categories. This includes a majority of residential, commercial, and industrial buildings.

- **Category III.** Buildings and structures that house a large number of people in one place, and buildings with occupants having limited ability to escape in the event of failure. Such buildings include theaters, elementary schools, and prisons. This category also includes structures associated with utilities and storage of hazardous materials.

- **Category IV.** Buildings and structures designated as essential facilities, such as hospitals and fire stations. This category also includes structures associated with storage of hazardous materials considered a danger to the public and buildings associated with utilities required to maintain the use of other buildings in this category.

Performance expectations for buildings vary widely depending on the type of hazard being resisted. Selection of the design event in the I-Codes is determined by the hazard type, the risk category of building, and the type of building damage expected. Selecting a higher risk category for most residential buildings should result in a higher final design wind pressure for design and should improve building performance in high-wind events. It can also result in additional freeboard in Zone V and Coastal A Zone if using ASCE 24 in flood design.

For *flood hazard design*, the building is divided into two distinct parts: the foundation and the main structure. For the foundation, standard methods of design target an essentially elastic response of the foundation for the design event such that little or no structural damage is expected. The main structure is designed to be constructed above the DFE to eliminate the need for designing it to resist flood loads. If flooding occurs at an elevation higher than the DFE, flood loads can be significant where flood waters impact solid walls (as opposed to open foundation elements). Additionally, a water level only a few inches above the minimum floor elevation can result in damage to walls and floors, and the loss of floor insulation, wiring, and ductwork. The IRC incorporates freeboard for houses in Zone V and Coastal A Zone, and the IBC

> **NOTE**
>
> Designing to only minimum code and regulatory requirements may result in designs based on different levels of risk for different hazards. The importance of each hazard level addressed by such requirements, and whether an acceptable level of residual risk remains, should therefore be carefully considered during the design process.

incorporates freeboard for buildings by virtue of using ASCE 24. Including freeboard in the building design provides a safety factor against damage to the main structure and its contents caused by flood elevations in excess of the design flood. While *codes and standards set minimum freeboard requirements*, a risk assessment may indicate the merits of incorporating additional freeboard above the minimum requirements (see Sections 6.2.1.3 and 6.3).

For *wind hazard design*, standard methods of design also target an essentially elastic response of the building structure for the design event (i.e., 700-year wind speed, 3-second gust per ASCE 7-10) such that little or no structural damage is expected. For wind speeds in excess of the design event, wind pressures increase predictably with wind velocity, and factors of safety associated with material resistances provide a margin against structural failure.

For *seismic hazard design*, life safety of the occupants is the primary focus rather than preventing any damage to the building. All portions of the building should be designed to resist the earthquake loads. Buildings are designed using the Maximum Considered Earthquake (i.e., 1 percent in 50 years) and include factors such as ground motion and peak ground acceleration. Adjustment factors are applied to design criteria based on the risk category for the building.

For *erosion hazard design* for bluff-top buildings, the ratio of soil strength to soil stresses is commonly used as the safety factor by geotechnical engineers when determining the risk of

> **NOTE**
>
> In the past, little thought was given to mitigation. Homeowners relied on insurance for replacement costs when a natural hazard event occurred, without regard to the inconvenience and disruption of their daily lives. Taking a mitigation approach can reduce these disruptions and inconveniences.

slope failures. The choice of a safety factor depends on the type and importance of bluff-top development, the bluff height, the nature of the potential bluff failure (e.g., deep rotational failure versus translational failure), and the acceptable level of risk associated with a bluff failure. Studies in the Great Lakes provide guidance for the selection of appropriate geotechnical safety factors (Valejo and Edil 1979, Chapman et al. 1996, and Terraprobe 1994).

6.2.1.2   Designing above Minimum Requirements and Preparing for Events That Exceed Design Events

In addition to incorporating factors of safety into design, homeowners, developers, and builders can make siting and design decisions that further manage risks by increasing the level of hazard resistance for the building. For example, hazard resistance can be improved by the following measures:

- A building can be sited further landward than the minimum distance specified by State or local setback requirements

- A building can be elevated above the level required by NFIP, State, and local requirements (refer to Section 6.2.1.3 for example)

- Supporting piles can be embedded deeper than required by State or local regulations

- Structural members and connections that exceed code requirements for gravity, uplift, and/or lateral forces can be used

- Improved roofing systems that provide greater resistance to wind than that required by code can be used

- Roof shapes (e.g., hip roofs) that reduce wind loads can be selected

- Openings (e.g., windows, doors) can be protected with permanent or temporary shutters or covers, whether or not such protection is required by code

- Enclosures below an elevated building can be eliminated or minimized

> **NOTE**
>
> While some coastal construction techniques have the combined effect of improving hazard resistance and energy efficiency, some design decisions make these considerations incompatible (see FEMA P-798, *Natural Hazards and Sustainability for Residential Buildings* [FEMA 2010]). Designers should discuss the implications and overall financial impacts of design decisions with homeowners so they can make an informed decision. The combination of insurance, maintenance, energy costs, and flood and wind resistance requires careful consideration and an understanding of the tradeoffs.

Incorporating above-code design can result in many benefits, such as reduced insurance premiums, reduced building maintenance, and potentially improved energy efficiency. These design decisions can sometimes offset the increased cost of constructing above the code minimums.

6.2.1.3   Role of Freeboard in Coastal Construction

The IRC and IBC (through ASCE 24) incorporate a minimum amount of freeboard. Including freeboard beyond that required by the NFIP and the building code should be seriously considered when designing for a homeowner with flooding risks. As of 2009, the IRC requires 1 foot of freeboard in Zone V and Coastal A

Zone. In most locations, designing for at least the freeboard requirements in ASCE 24, which requires more freeboard than the IRC in many cases, may establish the level of care expected of a design professional. Freeboard that exceeds the minimum NFIP requirements can be a valuable tool in maintaining NFIP compliance and lessening potential flood damage.

Some benefits of incorporating freeboard are:

CROSS REFERENCE

Section 7.5.2 includes a discussion of freeboard, BFE, and DFE.

- Allows lower flood insurance premiums

- Provides additional protection for floods exceeding the BFE

- Provides some contingency if future updates to FIRMs raise the BFE

- Helps account for changes within the SFHA that are not represented in the current FIRM or FIS

- Provides some contingency for surveying benchmarks that may have moved

- Provides some contingency for errors in the lowest floor elevation during construction without compromising the elevation above the BFE

- Provides some contingency for changes in water levels due to sea level change or subsidence

Even if a freeboard policy is not in force by the State or local jurisdiction, constructing a building to an elevation greater than the BFE reduces the homeowner's flood insurance premium. A FEMA report titled *Evaluation of the National Flood Insurance Program's Building Standards* (American Institutes for Research 2006) evaluates the benefits of freeboard. The report finds that freeboard is a cost-effective method for reducing risk in many instances and provides some guidance on the comparison of the percent increase in cost of construction with the reduced risk of flooding. Additionally, it evaluates the cost of construction for implementing freeboard and compares it to the flood insurance premium savings. A reevaluation of this study in December of 2009 validated that freeboard is still a cost-effective option in many coastal areas.

6.2.2 Managing Residual Risk through Insurance

Once all of the regulatory and physical risk reduction methods are incorporated into a building design, there will still be a level of residual risk to the building that must be assumed by homeowners. One way to minimize the financial exposure to the residual risk is through insurance. Insurance can be divided into a number of categories based on the type of hazard, and whether the insurance is private or purchased through a pool of other policy holders on a State or Federal level. While it is not the role of the designer to discuss insurance policies with an owner, it is important to understand the types of insurance available to an owner and the effect of building design decisions on various insurance programs. The following sections summarize of the types of hazard insurance and discuss how some design decisions can affect insurance premiums.

6.2.2.1   Types of Hazard Insurance

For houses in coastal areas, residual risks associated with flooding, high winds, and in some areas, earthquakes, are of particular concern. The financial risks can be mitigated through a variety of insurance mechanisms, including the NFIP, homeowners wind or earthquake insurance, insurance pools, and self-insurance plans.

National Flood Insurance Program

Federally backed flood insurance is available for both existing and new construction in communities that participate in the NFIP. To be insurable under the NFIP, a building must have a roof, have at least two walls, and be at least 50 percent above grade. Like homeowners insurance, flood insurance is obtained from private insurance companies. Flood insurance, because it is federally backed, is available for buildings in all coastal areas of participating communities, regardless of how high the flood hazard is. The following exceptions apply:

- Buildings constructed after October 1, 1982, that are entirely over water or seaward of mean high tide

- New construction, substantially improved, or substantially damaged buildings constructed after October 1, 1983, that are located on designated undeveloped coastal barriers included in the CBRS (see Section 5.1.1 of this Manual)

- Portions of boat houses located partially over water (e.g., the ceiling and roof over the area where boats are moored)

WARNING

Purchasing insurance is not a substitute for a properly designed and constructed building. Insurance is a way of reducing financial exposure to residual risk.

CROSS REFERENCE

For more information on hazard insurance, see Section 7.6.

COST CONSIDERATION

The NFIP places a cap on the amount of coverage for the building and its contents, which may not cover the entire cost of high value properties. Additional flood insurance will be required to insure losses above this limit.

The flood insurance rates for buildings in NFIP-participating communities vary according to the physical characteristics of the buildings, the date the buildings were constructed, and the magnitude of the flood hazard at the site of the buildings. The flood insurance premium for a building is based on the rate, standard per-policy fees, the amount of the deductible, applicable NFIP surcharges and discounts, and the amount of coverage obtained.

Wind Insurance

Homeowners insurance policies normally include coverage for wind. However, insurance companies that issue homeowner policies occasionally deny wind coverage to buildings in areas where the risks from these hazards are high, especially in coastal areas subject to a significant hurricane or typhoon risk. At the time of publication of this Manual, underwriting associations,

NOTE

The Florida Division of Emergency Management has an online insurance savings calculator that estimates wind insurance savings for wind mitigation design in new construction and retrofits. The calculator is available at http://floridadisaster.org/mitdb.

or "pools," are a last resort for homeowners who need wind coverage but cannot obtain it from private companies. Seven States have beach and wind insurance plans: Alabama, Florida, Louisiana, Mississippi, North Carolina, South Carolina, and Texas. Georgia and New York provide this kind of coverage for windstorms and hail in certain coastal communities through other property pools. In addition, New Jersey operates the Windstorm Market Assistance Program (Wind-MAP; http://www.njiua.org) to help residents in coastal communities find homeowners insurance in the voluntary market. When Wind-MAP does not identify an insurance carrier for a homeowner, the homeowner may apply to the New Jersey Insurance Underwriting Association, known as the FAIR Plan, for a perils-only policy.

> **WARNING**
>
> Hurricanes cause damage through wind and flooding; however, flood insurance policies only cover flood damage, and wind insurance policies only cover damage from wind and wind-driven rain. For more comprehensive insurance protection, property owners should invest in both flood and wind insurance.

Earthquake Insurance

A standard homeowners insurance policy can often be modified through an endorsement to include earthquake coverage. However, like wind coverage, earthquake coverage may not be available in areas where the earthquake risk is high. Moreover, deductibles and rates for earthquake coverage (of typical coastal residential buildings) are usually much higher than those for flood, wind, and other hazard insurance.

Self-Insurance

Where wind and earthquake insurance coverage is not available from private companies or insurance pools—or where homeowners choose to forego available insurance—owners with sufficient financial reserves may be able to assume complete financial responsibility for the risks not offset through siting, design, construction, and maintenance (i.e., self-insure). Homeowners who contemplate self-insurance must understand the true level of risk they are assuming.

6.2.2.2 Savings, Premium, and Penalties

Design and siting decisions can often have a dramatic effect on both flood and wind insurance premiums. The primary benefit of the guidance in this Manual is the reduction of damage, disruption, and risk to the client. However, the reduction of insurance costs is a secondary benefit. Siting a building farther from the coastline could result in moving a building from Zone V into Zone A, thereby reducing premiums. Additionally, the height of the structure can affect flood insurance premiums. Raising the first floor elevation above the BFE (adding freeboard) reduces premiums in all flood zones.

Some design decisions increase, rather than decrease, insurance premiums. For instance, while the NFIP allows for enclosures below the lowest floor, their presence may increase flood insurance premiums. Breakaway walls and floor systems

> **COST CONSIDERATION**
>
> Constructing enclosures can have significant cost implications. This Manual recommends the use of insect screening or open wood lattice instead of solid enclosures beneath elevated residential buildings. See also Section 2.3.5 of this Manual.

elevated off the ground can raise premiums. Although these are allowed by the program, these types of design elements should be considered carefully and discussed with homeowners in light of their overall long-term cost implications.

In some States, building a house stronger than required by code results in reduced wind insurance premiums. For example, Florida requires insurance companies to offer discounts or credits for design and construction techniques that reduce damage and loss in windstorms. Stronger roofs and wall systems and improved connections may reduce premiums. Conversely, the addition of large overhangs and other building elements that increase the building's wind exposure can increase premiums. Building a structure stronger than the minimum code can have the dual benefit of reducing insurance premiums and decreasing damage during a flood or wind event.

> **WARNING**
>
> Improper construction of enclosures below elevated residential buildings in Zone V and post-construction conversion of enclosed space to habitable use (in Zone A and Zone V) are common compliance violations of the NFIP. For more guidance on enclosures, see Section 2.3.5 of this Manual.

6.3   Communicating Risk to Clients

Many homeowners may not be aware of the hazards that could affect their property and may not understand the risk they assume through their design decisions. Communicating risk to homeowners in a variety of ways, both technical and non-technical, is important so they understand the benefits and drawbacks of decisions they make. Designers should communicate how design decisions and material selections (as discussed in Volume II) can reduce risk, and the mitigation of residual risk through insurance.

It is important for homeowners to understand how the choices they make in designing their home could potentially reduce its risk of being damaged or destroyed by natural hazards. Designers need to be familiar with the potential risks for the property and be prepared to suggest design measures that not only meet the needs and tastes of homeowners, but that also provide protection from hazard impacts. In addition, design choices that have implications for building performance during a hazard event and on insurance premiums should be discussed clearly with the homeowner.

Although the effects of natural hazards can be reduced through thoughtful design and construction, *homeowners should understand that there will always be residual risk* from coastal hazards as long as they choose to build in a coastal environment. Proper design elements can mitigate some of those risks, but there is no way to completely eliminate residual risk in coastal areas. As described in this chapter, mitigating natural hazard risk in a coastal environment entails implementing a series of risk reduction methods, such as physical risk reduction and risk management through insurance. While some level of residual risk will remain, owners can use these tools to protect themselves and their investments.

Homeowners often misunderstand their risk; therefore, risk communication is critical to help them understand the risk that they assume. Designers are often tasked with explaining complicated risk concepts to homeowners. The discussion of risk with a homeowner can be difficult. It is important to find methods to convey the natural hazard risks for a site and how those risks may be addressed in the design process. The following discussion and examples are provided for designers to use with their clients. These examples use comparisons to other hazards, graphics, and monetary comparisons to provide alternatives to annual probabilities and recurrence intervals.

6.3.1   Misconceptions about the 100-Year Flood Event

Homeowners commonly misunderstand the *1-percent-annual-chance flood, often called the 100-year flood*. There is a 1 percent chance each year of the occurrence of a flood that equals or exceeds the BFE. By contrast, the chance of burglary in 2005 was only 0.6 percent nationwide, but homeowners are concerned enough by this threat that they use security systems and buy homeowners insurance to cover their belongings. Many homeowners believe that being in the 1-percent-annual-chance floodplain means that there is only a 1 percent chance of ever being flooded, which they deem a very small risk. Another misconception is that the "100-year" flood only happens once every 100 years. Unfortunately, these misconceptions result in a gross underestimation of their flood risk. In reality, a residential building within the SFHA has a 26 percent chance of being damaged by a flood over the course of a 30-year mortgage, compared to a 10 percent chance of fire or 17 percent chance of burglary.

6.3.2   Misconceptions about Levee Protection

Another common misconception involves levee protection. *Many homeowners behind a levee believe that the levee will protect their property from flood so they believe they are not at risk*. Since each levee is constructed to provide protection against a specific flood frequency, the level of protection must be identified before the risk can be identified. Owners and designers must understand that because levees are only

CROSS REFERENCE

Section 2.3.2 discusses building behind a levee.

designed to withstand certain storm event recurrence intervals, they may fail when a greater-than-design event occurs. Additional risk factors include the age of the levee and whether the level of protection provided by it may have changed over time. Designers must also understand that levees may have been designed for a specific level of protection, but if flood data changes over time due to an improved understanding of flood modeling, the current level of protection may be less than the designed level of protection. If a levee should fail or is overtopped, the properties behind the levee will be damaged by flooding, which could be as damaging as if there were no levee there at all. Therefore, even in levee-protected areas, homeowners need to be aware of the risk and should consider elevation and other mitigation techniques to minimize their flood risk.

EXAMPLE: ELEVATING ABOVE THE MINIMUM CONSTRUCTION ELEVATION

Consider the following example of how just one decision made by the designer, builder, or homeowner can affect risk. Local floodplain management requirements consistent with NFIP regulations require that any building constructed in Zone V be elevated so that the bottom of the lowest horizontal structural member is at or above the BFE (1-percent-annual-chance flood elevation, including wave effects). Meeting this requirement should protect the elevated portion of the building from the 1-percent-annual-chance and lesser floods. However, the elevated part of the building is still vulnerable to floods of greater magnitude. As shown in Table 6-1, the probability that the building will be subjected to a flood greater than the 1-percent-annual-chance flood during a period of 30 years is 26 percent. But during the same 30-year period, the probability of a 0.2-percent-annual-chance ("500-year") or greater flood is only 7 percent. Therefore, raising the lowest horizontal structural member to the elevation of the 0.2-percent-annual-chance flood would significantly reduce the building's vulnerability to flooding and reduce insurance premiums. If elevating to the level of the 0.2-percent-annual-chance flood is not possible because of cost or other considerations, elevating by some lesser amount above the BFE will still reduce the risk.

> **CROSS REFERENCE**
>
> Sections 6.2.1.3 and 6.2.2 provide some discussion on how raising the lowest horizontal structural member to the elevation of the 0.2-percent-annual-chance flood instead of the BFE would provide benefits by reducing both the physical risk to the structure and the insurance premiums.

Illustration A on the next page shows the percent chance over a 30-year period of houses being flooded. The left side of the illustration reflects houses constructed to the BFE, while the right side reflects houses constructed to an elevation above the BFE, the 0.2-percent-annual-chance ("500-year") flood elevation. Explain to the homeowner that the number of flooded houses shown is the percent of houses that would be potentially flooded over the next 30 years in each condition. Constructing to the 0.2-percent-annual-chance flood elevation reduces both physical risk and insurance cost. Illustration B shows the potential cost savings over a 30-year period for a house constructed to the BFE and a house constructed to the 0.2-percent-annual-chance flood elevation. For the purposes of calculating costs, the difference in elevation between BFE and the 0.2-percent-annual-chance flood in this example is 3 feet. The difference in elevation between the BFE and the 0.2-percent-annual-chance flood actually varies by location.

After a quick overview of the illustrations, most homeowners will understand how elevating the building higher than the BFE can result in significantly lower chances of the house experiencing flooding over the next 30 years. Once they understand the advantages of elevating a house higher than the minimum, they can be shown that while constructing the house higher will result in increased construction costs, it will also result in reduced flood insurance premiums. The designer can further explain that these reduced flood insurance premiums will quickly offset the increased construction costs. In this example, spending an additional $12,000 in construction costs to build the house 3 feet above the BFE will save the homeowner $151,710 in premiums over a 30-year mortgage period (for a total savings of $139,710). Designers can use illustrations such as these or other such comparisons to explain exposure to natural hazards, risk, and reasons for making design decisions.

EXAMPLE: ELEVATING ABOVE THE MINIMUM CONSTRUCTION ELEVATION
(concluded)

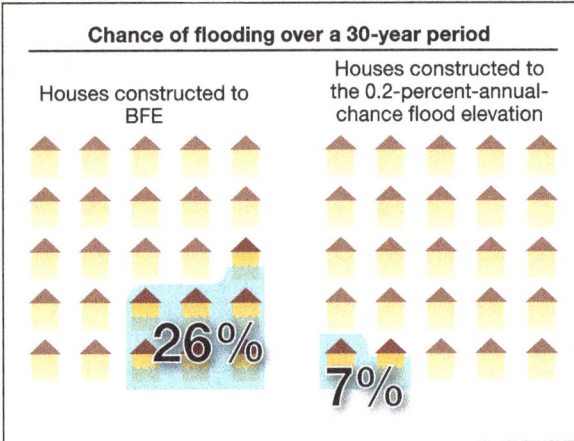

Chance of flooding over a 30-year period

Houses constructed to BFE

Houses constructed to the 0.2-percent-annual-chance flood elevation

26%

7%

Illustration A:
Comparison of the percent chance of houses being flooded over a 30-year period after being elevated to the BFE (left) and the 0.2-percent-annual-chance flood elevation (right)

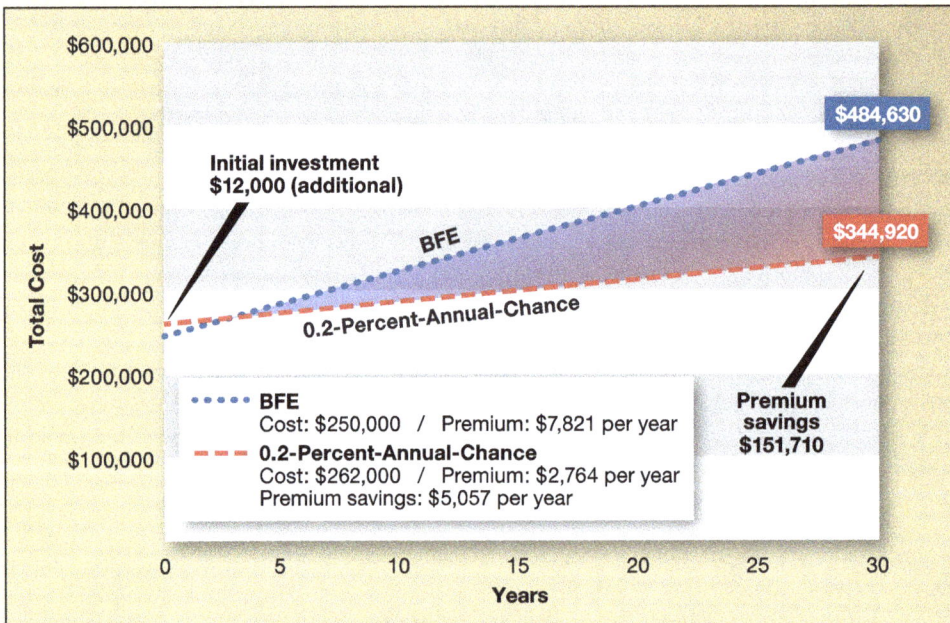

Initial investment $12,000 (additional)

BFE

0.2-Percent-Annual-Chance

$484,630

$344,920

Premium savings $151,710

····· **BFE**
Cost: $250,000 / Premium: $7,821 per year

--- **0.2-Percent-Annual-Chance**
Cost: $262,000 / Premium: $2,764 per year
Premium savings: $5,057 per year

Total Cost / Years

Illustration B:
Comparison of the total cost over a 30-year period for a house elevated to the BFE (dotted line) and a house elevated to the 0.2-annual-chance flood elevation (dashed line)

Note:
This example includes the cost of adding 3 feet of freeboard above the BFE, elevating the house to the 0.2-percent-annual-chance flood elevation. The difference in elevation between the BFE and the 0.2-percent-annual-chance flood actually varies by location.

Example premiums calculated using the NFIP Flood Insurance Manual, May 1, 2011, for a Zone V structure free of obstructions. Premiums include building ($250,000), contents ($100,000), and associated fees including Increased Cost of Compliance.

6.4   References

American Institutes for Research. 2006. *Evaluation of the National Flood Insurance Program's Building Standards.* October.

American Society of Civil Engineers (ASCE). 2005. *Flood Resistant Design and Construction.* ASCE Standard ASCE 24-05.

ASCE. 2010. *Minimum Design Loads of Buildings and Other Structures.* ASCE Standard ASCE 7-10.

Chapman, J. A.; T. B. Edil; and D. M. Mickelson. 1996. *Effectiveness of Analysis Methods for Predicting Long Term Slope Stability on the Lake Michigan Shoreline.* Final report of the Lake Michigan, Wisconsin, Shoreline Erosion and Bluff Stability Study. December.

FEMA (Federal Emergency Management Agency). 2010. *Natural Hazards and Sustainability for Residential Buildings.* FEMA P-798. September.

FEMA. 2011. *Flood Insurance Manual.* May.

Potter, K. W. 1992. "Estimating the Probability Distribution of Annual Maximum Levels on the Great Lakes." *Journal of Great Lakes Research, International Association of Great Lakes Research.* (18)1, pp. 229–235.

Terraprobe. 1994. *Geotechnical Principles for Stable Slopes. Great Lakes-St. Lawrence River Shoreline Policy.* Prepared for the Ontario Ministry of Natural Resources. Terraprobe, Ltd., Brampton, Ontario. March.

Valejo, L. E.; T. B. Edil. 1979. "Design Charts for Development and Stability of Evolving Slopes." *Journal of Civil Engineering Design.* Marcel Dekker, Inc., publisher, 1(3), pp. 232–252.

Acronyms

A

AF&PA	American Forest & Paper Association
AIA	American Institute of Architects
AISI	American Iron and Steel Institute
APA	American Planning Association
ASCE	American Society of Civil Engineers
ASD	allowable-stress design
ASFPM	Association of State Floodplain Managers
ASLA	American Society of Landscape Architects

B

BCEGS	Building Code Effectiveness Grading Schedule
BFE	base flood elevation
BOCA	Building Officials and Code Administrators International, Inc.
BPAT	Building Performance Assessment Team

C

CBIA	Coastal Barrier Improvement Act
CBRA	Coastal Barrier Resources Act
CBRS	Coastal Barrier Resources System

CFR	Code of Federal Regulations
CRS	Community Rating System
CZM	Coastal Zone Management
CZMA	Coastal Zone Management Act

DFE	design flood elevation
DFIRM	Digital Flood Insurance Rate Map
DOI	Department of the Interior

EF	Enhanced Fujita
EHP	Earthquake Hazards Program
ENSO	El Nino/La Nina-Southern Oscillation

FBC	Florida Building Code
FEMA	Federal Emergency Management Agency
FIRM	Flood Insurance Rate Map
FIS	Flood Insurance Study

GSA	General Services Administration

HUD	Department of Housing and Urban Development

IBC	International Building Code
IBHS	Insurance Institute for Business and Home Safety
ICBO	International Conference of Building Officials
ICC	International Code Council
IEBC	International Existing Building Code
IFC	International Fire Code
IFGC	International Fuel Gas Code
IMC	International Mechanical Code
IPC	International Plumbing Code
IPCC	Intergovernmental Panel on Climate Change
IPSDC	International Private Sewage Disposal Code
IRC	International Residential Code
ISO	Insurance Services Office

LiMWA	Limit of Moderate Wave Action

MAT	Mitigation Assessment Team
MiWA	Minimal Wave Action
MMI	Modified Mercalli Intensity
MoWA	Moderate Wave Action

NAHB	National Association of Home Builders

NAVD	North American Vertical Datum
NBC	National Building Code
NEHRP	National Earthquake Hazards Reduction Program
NFIP	National Flood Insurance Program
NFPA	National Fire Protection Association
NGVD	National Geodetic Vertical Datum
NIST	National Institute of Standards and Technology
NOAA	National Oceanic and Atmospheric Administration
NRCA	National Roofing Contractors Association
NRCS	Natural Resources Conservation Service
NSPE	National Society of Professional Engineers
NWS	National Weather Service
OCRM	Office of Ocean and Coastal Resource Management
OPA	Otherwise Protected Area
SBC	Standard Building Code
SBCCI	Southern Building Code Congress International
SDE	Substantial Damage Estimator
SFHA	Special Flood Hazard Area
SFIP	Standard Flood Insurance Policy
SI/SD	Substantial Improvement/Substantial Damage
SSHWS	Saffir-Simpson Hurricane Wind Scale

UBC   Uniform Building Code

USACE   U.S. Army Corps of Engineers

USGS   U.S. Geological Survey

Wind-MAP   Windstorm Market Assistance Program (New Jersey)

Glossary

0-9

100-year flood – See *Base flood*.

500-year flood – Flood that has as 0.2-percent probability of being equaled or exceeded in any given year.

A

Acceptable level of risk – The level of risk judged by the building owner and designer to be appropriate for a particular building.

Adjacent grade – Elevation of the natural or graded ground surface, or structural fill, abutting the walls of a building. See also *Highest adjacent grade* and *Lowest adjacent grade*.

Angle of internal friction (soil) – A measure of the soil's ability to resist shear forces without failure.

Appurtenant structure – Under the National Flood Insurance Program, an "appurtenant structure" is "a structure which is on the same parcel of property as the principal structure to be insured and the use of which is incidental to the use of the principal structure."

B

Barrier island – A long, narrow sand island parallel to the mainland that protects the coast from erosion.

Base flood – Flood that has as 1-percent probability of being equaled or exceeded in any given year. Also known as the 100-year flood.

Base Flood Elevation (BFE) – The water surface elevation resulting from a flood that has a 1 percent chance of equaling or exceeding that level in any given year. Elevation of the base flood in relation to a specified datum, such as the National Geodetic Vertical Datum or the North American Vertical Datum. The Base Flood Elevation is the basis of the insurance and floodplain management requirements of the National Flood Insurance Program.

Basement – Under the National Flood Insurance Program, any area of a building having its floor subgrade on all sides. (Note: What is typically referred to as a "walkout basement," which has a floor that is at or above grade on at least one side, is not considered a basement under the National Flood Insurance Program.)

Beach nourishment – A project type that typically involve dredging or excavating hundreds of thousands to millions of cubic yards of sediment, and placing it along the shoreline.

Bearing capacity (soils) – A measure of the ability of soil to support gravity loads without soil failure or excessive settlement.

Berm – Horizontal portion of the backshore beach formed by sediments deposited by waves.

Best Practices – Techniques that exceed the minimum requirements of model building codes; design and construction standards; or Federal, State, and local regulations.

Breakaway wall – Under the National Flood Insurance Program, a wall that is not part of the structural support of the building and is intended through its design and construction to collapse under specific lateral loading forces without causing damage to the elevated portion of the building or supporting foundation system. Breakaway walls are required by the National Flood Insurance Program regulations for any enclosures constructed below the Base Flood Elevation beneath elevated buildings in Coastal High Hazard Areas (also referred to as Zone V). In addition, breakaway walls are recommended in areas where flood waters flow at high velocities or contain ice or other debris.

Building code – Regulations adopted by local governments that establish standards for construction, modification, and repair of buildings and other structures.

Building use – What occupants will do in the building. The intended use of the building will affect its layout, form, and function.

Building envelope – Cladding, roofing, exterior walls, glazing, door assemblies, window assemblies, skylight assemblies, and other components enclosing the building.

Building systems – Exposed structural, window, or roof systems.

Built-up roof covering – Two or more layers of felt cemented together and surfaced with a cap sheet, mineral aggregate, smooth coating, or similar surfacing material.

Bulkhead – Wall or other structure, often of wood, steel, stone, or concrete, designed to retain or prevent sliding or erosion of the land. Occasionally, bulkheads are used to protect against wave action.

Cladding – Exterior surface of the building envelope that is directly loaded by the wind.

Closed foundation – A foundation that does not allow water to pass easily through the foundation elements below an elevated building. Examples of closed foundations include crawlspace foundations and stem wall foundations, which are usually filled with compacted soil, slab-on-grade foundations, and continuous perimeter foundation walls.

Coastal A Zone – The portion of the coastal SFHA referenced by building codes and standards, where base flood wave heights are between 1.5 and 3 feet, and where wave characteristics are deemed sufficient to damage many NFIP-compliant structures on shallow or solid wall foundations.

Coastal barrier – Depositional geologic feature such as a bay barrier, tombolo, barrier spit, or barrier island that consists of unconsolidated sedimentary materials; is subject to wave, tidal, and wind energies; and protects landward aquatic habitats from direct wave attack.

Coastal Barrier Resources Act of 1982 (CBRA) – Act (Public Law 97-348) that established the Coastal Barrier Resources System (CBRS). The act prohibits the provision of new flood insurance coverage on or after October 1, 1983, for any new construction or substantial improvements of structures located on any designated undeveloped coastal barrier within the CBRS. The CBRS was expanded by the Coastal Barrier Improvement Act of 1991. The date on which an area is added to the CBRS is the date of CBRS designation for that area.

Coastal flood hazard area – An area subject to inundation by storm surge and, in some instances, wave action caused by storms or seismic forces. Usually along an open coast, bay, or inlet.

Coastal geology – The origin, structure, and characteristics of the rocks and sediments that make up the coastal region.

Coastal High Hazard Area – Under the National Flood Insurance Program, an area of special flood hazard extending from offshore to the inland limit of a primary frontal dune along an open coast and any other area subject to high-velocity wave action from storms or seismic sources. On a Flood Insurance Rate Map, the Coastal High Hazard Area is designated Zone V, VE, or V1-V30. These zones designate areas subject to inundation by the base flood, where wave heights or wave runup depths are 3.0 feet or higher.

Coastal processes – The physical processes that act upon and shape the coastline. These processes, which influence the configuration, orientation, and movement of the coast, include tides and fluctuating water levels, waves, currents, and winds.

Coastal sediment budget – The quantification of the amounts and rates of sediment transport, erosion, and deposition within a defined region.

Coastal Special Flood Hazard Area – The portion of the Special Flood Hazard Area where the source of flooding is coastal surge or inundation. It includes Zone VE and Coastal A Zone.

Code official – Officer or other designated authority charged with the administration and enforcement of the code, or a duly authorized representative, such as a building, zoning, planning, or floodplain management official.

Column foundation – Foundation consisting of vertical support members with a height-to-least-lateral-dimension ratio greater than three. Columns are set in holes and backfilled with compacted material. They are usually made of concrete or masonry and often must be braced. Columns are sometimes known as posts, particularly if they are made of wood.

Components and Cladding (C&C) – American Society of Civil Engineers (ASCE) 7-10 defines C&C as "... elements of the building envelope that do not qualify as part of the MWFRS [Main Wind Force Resisting System]." These elements include roof sheathing, roof coverings, exterior siding, windows, doors, soffits, fascia, and chimneys.

Conditions Greater than Design Conditions – Design loads and conditions are based on some probability of exceedance, and it is always possible that design loads and conditions can be exceeded. Designers can anticipate this and modify their initial design to better accommodate higher forces and more extreme conditions. The benefits of doing so often exceed the costs of building higher and stronger.

Connector – Mechanical device for securing two or more pieces, parts, or members together, including anchors, wall ties, and fasteners.

Consequence – Both the short- and long-term effects of an event for the building. See *Risk.*

Constructability – Ultimately, designs will only be successful if they can be implemented by contractors. Complex designs with many custom details may be difficult to construct and could lead to a variety of problems, both during construction and once the building is occupied.

Continuous load paths – The structural condition required to resist loads acting on a building. The continuous load path starts at the point or surface where loads are applied, moves through the building, continues through the foundation, and terminates where the loads are transferred to the soils that support the building.

Corrosion-resistant metal – Any nonferrous metal or any metal having an unbroken surfacing of nonferrous metal, or steel with not less than 10 percent chromium or with not less than 0.20 percent copper.

Dead load – Weight of all materials of construction incorporated into the building, including but not limited to walls, floors, roofs, ceilings, stairways, built-in partitions, finishes, cladding, and other similarly incorporated architectural and structural items and fixed service equipment. See also *Loads.*

Debris – Solid objects or masses carried by or floating on the surface of moving water.

Debris impact loads – Loads imposed on a structure by the impact of floodborne debris. These loads are often sudden and large. Though difficult to predict, debris impact loads must be considered when structures are designed and constructed. See also *Loads.*

Deck – Exterior floor supported on at least two opposing sides by an adjacent structure and/or posts, piers, or other independent supports.

Design event – The minimum code-required event (for natural hazards, such as flood, wind, and earthquake) and associated loads that the structure must be designed to resist.

Design flood – The greater of either (1) the base flood or (2) the flood associated with the flood hazard area depicted on a community's flood hazard map, or otherwise legally designated.

Design Flood Elevation (DFE) – Elevation of the design flood, or the flood protection elevation required by a community, including wave effects, relative to the National Geodetic Vertical Datum, North American Vertical Datum, or other datum. The DFE is the locally adopted regulatory flood elevation. If a community regulates to minimum National Flood Insurance Program (NFIP) requirements, the

DFE is identical to the Base Flood Elevation (BFE). If a community chooses to exceed minimum NFIP requirements, the DFE exceeds the BFE.

Design flood protection depth – Vertical distance between the eroded ground elevation and the Design Flood Elevation.

Design stillwater flood depth – Vertical distance between the eroded ground elevation and the design stillwater flood elevation.

Design stillwater flood elevation – Stillwater elevation associated with the design flood, excluding wave effects, relative to the National Geodetic Vertical Datum, North American Vertical Datum, or other datum.

Development – Under the National Flood Insurance Program, any manmade change to improved or unimproved real estate, including but not limited to buildings or other structures, mining, dredging, filling, grading, paving, excavation, or drilling operations or storage of equipment or materials.

Dry floodproofing – A flood retrofitting technique in which the portion of a structure below the flood protection level (walls and other exterior components) is sealed to be impermeable to the passage of floodwaters.

Dune – See *Frontal dune* and *Primary frontal dune*.

Dune toe – Junction of the gentle slope seaward of the dune and the dune face, which is marked by a slope of 1 on 10 or steeper.

Effective Flood Insurance Rate Map – See *Flood Insurance Rate Map*.

Elevation – Raising a structure to prevent floodwaters from reaching damageable portions.

Enclosure – The portion of an elevated building below the lowest floor that is partially or fully shut in by rigid walls.

Encroachment – The placement of an object in a floodplain that hinders the passage of water or otherwise affects the flood flows.

Erodible soil – Soil subject to wearing away and movement due to the effects of wind, water, or other geological processes during a flood or storm or over a period of years.

Erosion – Under the National Flood Insurance Program, the process of the gradual wearing away of land masses.

Erosion analysis – Analysis of the short- and long-term erosion potential of soil or strata, including the effects of flooding or storm surge, moving water, wave action, and the interaction of water and structural components.

Exterior-mounted mechanical equipment – Includes, but is not limited to, exhaust fans, vent hoods, air conditioning units, duct work, pool motors, and well pumps.

Federal Emergency Management Agency (FEMA) – Independent agency created in 1979 to provide a single point of accountability for all Federal activities related to disaster mitigation and emergency preparedness, response, and recovery. FEMA administers the National Flood Insurance Program.

Federal Insurance and Mitigation Administration (FIMA) – The component of the Federal Emergency Management Agency directly responsible for administering the flood insurance aspects of the National Flood Insurance Program as well as a range of programs designed to reduce future losses to homes, businesses, schools, public buildings, and critical facilities from floods, earthquakes, tornadoes, and other natural disasters.

Fill – Material such as soil, gravel, or crushed stone placed in an area to increase ground elevations or change soil properties. See also *Structural fill*.

Flood – Under the National Flood Insurance Program, either a general and temporary condition or partial or complete inundation of normally dry land areas from:

(1) the overflow of inland or tidal waters;

(2) the unusual and rapid accumulation or runoff of surface waters from any source;

(3) mudslides (i.e., mudflows) that are proximately caused by flooding as defined in (2) and are akin to a river of liquid and flowing mud on the surfaces of normally dry land areas, as when the earth is carried by a current of water and deposited along the path of the current; or

(4) the collapse or subsidence of land along the shore of a lake or other body of water as a result of erosion or undermining caused by waves or currents of water exceeding anticipated cyclical levels or suddenly caused by an unusually high water level in a natural body of water, accompanied by a severe storm, or by an unanticipated force of nature, such as flash flood or abnormal tidal surge, or by some similarly unusual and unforeseeable event which results in flooding as defined in (1), above.

Flood-damage-resistant material – Any construction material capable of withstanding direct and prolonged contact (i.e., at least 72 hours) with flood waters without suffering significant damage (i.e., damage that requires more than cleanup or low-cost cosmetic repair, such as painting).

Flood elevation – Height of the water surface above an established elevation datum such as the National Geodetic Vertical Datum, North American Vertical Datum, or mean sea level.

Flood hazard area – The greater of the following: (1) the area of special flood hazard, as defined under the National Flood Insurance Program, or (2) the area designated as a flood hazard area on a community's legally adopted flood hazard map, or otherwise legally designated.

Flood insurance – Insurance coverage provided under the National Flood Insurance Program.

Flood Insurance Rate Map (FIRM) – Under the National Flood Insurance Program, an official map of a community, on which the Federal Emergency Management Agency has delineated both the special hazard areas and the risk premium zones applicable to the community. (Note: The latest FIRM issued for a community is referred to as the "effective FIRM" for that community.)

Flood Insurance Study (FIS) – Under the National Flood Insurance Program, an examination, evaluation, and determination of flood hazards and, if appropriate, corresponding water surface elevations, or an examination, evaluation, and determination of mudslide (i.e., mudflow) and flood-related erosion hazards in a community or communities. (Note: The National Flood Insurance Program regulations refer to Flood Insurance Studies as "flood elevation studies.")

Flood-related erosion area or flood-related erosion prone area – A land area adjoining the shore of a lake or other body of water, which due to the composition of the shoreline or bank and high water levels or wind-driven currents, is likely to suffer flood-related erosion.

Flooding – See *Flood*.

Floodplain – Under the National Flood Insurance Program, any land area susceptible to being inundated by water from any source. See also *Flood*.

Floodplain management – Operation of an overall program of corrective and preventive measures for reducing flood damage, including but not limited to emergency preparedness plans, flood control works, and floodplain management regulations.

Floodplain management regulations – Under the National Flood Insurance Program, zoning ordinances, subdivision regulations, building codes, health regulations, special purpose ordinances (such as floodplain ordinance, grading ordinance, and erosion control ordinance), and other applications of police power. The term describes State or local regulations, in any combination thereof, that promulgate standards for the purpose of flood damage prevention and reduction.

Floodwall – A flood retrofitting technique that consists of engineered barriers designed to keep floodwaters from coming into contact with the structure.

Footing – Enlarged base of a foundation wall, pier, post, or column designed to spread the load of the structure so that it does not exceed the soil bearing capacity.

Footprint – Land area occupied by a structure.

Freeboard – Under the National Flood Insurance Program, a factor of safety, usually expressed in feet above a flood level, for the purposes of floodplain management. Freeboard is intended to compensate for the many unknown factors that could contribute to flood heights greater than the heights calculated for a selected size flood and floodway conditions, such as the hydrological effect of urbanization of the watershed. Freeboard is additional height incorporated into the Design Flood Elevation, and may be required by State or local regulations or be desired by a property owner.

Frontal dune – Ridge or mound of unconsolidated sandy soil extending continuously alongshore landward of the sand beach and defined by relatively steep slopes abutting markedly flatter and lower regions on each side.

Frontal dune reservoir – Dune cross-section above 100-year stillwater level and seaward of dune peak.

Gabion – Rock-filled cage made of wire or metal that is placed on slopes or embankments to protect them from erosion caused by flowing or fast-moving water.

Geomorphology – The origin, structure, and characteristics of the rocks and sediments that make up the coastal region.

Glazing – Glass or transparent or translucent plastic sheet in windows, doors, skylights, and shutters.

Grade beam – Section of a concrete slab that is thicker than the slab and acts as a footing to provide stability, often under load-bearing or critical structural walls. Grade beams are occasionally installed to provide lateral support for vertical foundation members where they enter the ground.

High-velocity wave action – Condition in which wave heights or wave runup depths are 3.0 feet or higher.

Highest adjacent grade – Elevation of the highest natural or regraded ground surface, or structural fill, that abuts the walls of a building.

Hurricane – Tropical cyclone, formed in the atmosphere over warm ocean areas, in which wind speeds reach 74 miles per hour or more and blow in a large spiral around a relatively calm center or "eye." Hurricane circulation is counter-clockwise in the northern hemisphere and clockwise in the southern hemisphere.

Hurricane clip or strap – Structural connector, usually metal, used to tie roof, wall, floor, and foundation members together so that they resist wind forces.

Hurricane-prone region – In the United States and its territories, hurricane-prone regions are defined by The American Society of Civil Engineers (ASCE) 7-10 as: (1) The U.S. Atlantic Ocean and Gulf of Mexico coasts where the basic wind speed for Risk Category II buildings is greater than 115 mph, and (2) Hawaii, Puerto Rico, Guam, the Virgin Islands, and American Samoa.

Hydrodynamic loads – Loads imposed on an object, such as a building, by water flowing against and around it. Among these loads are positive frontal pressure against the structure, drag effect along the sides, and negative pressure on the downstream side.

Hydrostatic loads – Loads imposed on a surface, such as a wall or floor slab, by a standing mass of water. The water pressure increases with the square of the water depth.

Initial costs – Include property evaluation, acquisition, permitting, design, and construction.

Interior mechanical equipment – Includes, but is not limited to, furnaces, boilers, water heaters, and distribution ductwork.

Jetting (of piles) – Use of a high-pressure stream of water to embed a pile in sandy soil. See also *Pile foundation*.

Jetty – Wall built from the shore out into the water to restrain currents or protect a structure.

Joist – Any of the parallel structural members of a floor system that support, and are usually immediately beneath, the floor.

Lacustrine flood hazard area – Area subject to inundation from lakes.

Landslide – Occurs when slopes become unstable and loose material slides or flows under the influence of gravity. Often, landslides are triggered by other events such as erosion at the toe of a steep slope, earthquakes, floods, or heavy rains, but can be worsened by human actions such as destruction of vegetation or uncontrolled pedestrian access on steep slopes.

Levee – Typically a compacted earthen structure that blocks floodwaters from coming into contact with the structure, a levee is a manmade structure built parallel to a waterway to contain, control, or divert the flow of water. A levee system may also include concrete or steel floodwalls, fixed or operable floodgates and other closure structures, pump stations for rainwater drainage, and other elements, all of which must perform as designed to prevent failure.

Limit of Moderate Wave Action (LiMWA) – A line indicating the limit of the 1.5-foot wave height during the base flood. FEMA requires new flood studies in coastal areas to delineate the LiMWA.

Littoral drift – Movement of sand by littoral (longshore) currents in a direction parallel to the beach along the shore.

Live loads – Loads produced by the use and occupancy of the building or other structure. Live loads do not include construction or environmental loads such as wind load, snow load, rain load, earthquake load, flood load, or dead load. See also *Loads*.

Load-bearing wall – Wall that supports any vertical load in addition to its own weight. See also *Non-load-bearing wall*.

Loads – Forces or other actions that result from the weight of all building materials, occupants and their possessions, environmental effects, differential movement, and restrained dimensional changes. Loads can be either permanent or variable. Permanent loads rarely vary over time or are of small magnitude. All other loads are variable loads.

Location – The location of the building determines the nature and intensity of hazards to which the building will be exposed, loads and conditions that the building must withstand, and building regulations that must be satisfied. See also *Siting*.

Long-term costs – Include preventive maintenance and repair and replacement of deteriorated or damaged building components. A hazard-resistant design can result in lower long-term costs by preventing or reducing losses from natural hazards events.

Lowest adjacent grade (LAG) – Elevation of the lowest natural or regraded ground surface, or structural fill, that abuts the walls of a building. See also *Highest adjacent grade*.

Lowest floor – Under the National Flood Insurance Program (NFIP), "lowest floor" of a building includes the floor of a basement. The NFIP regulations define a basement as "... any area of a building having its floor subgrade (below ground level) on all sides." For insurance rating purposes, this definition applies even when the subgrade floor is not enclosed by full-height walls.

Lowest horizontal structural member – In an elevated building, the lowest beam, joist, or other horizontal member that supports the building. Grade beams installed to support vertical foundation members where they enter the ground are not considered lowest horizontal structural members.

Main Wind Force Resisting System (MWFRS) – Consists of the foundation; floor supports (e.g., joists, beams); columns; roof raters or trusses; and bracing, walls, and diaphragms that assist in transferring loads. The American Society of Civil Engineers (ASCE) 7-10 defines the MWFRS as "... an assemblage of structural elements assigned to provide support and stability for the overall structure."

Manufactured home – Under the National Flood Insurance Program, a structure, transportable in one or more sections, built on a permanent chassis and designed for use with or without a permanent foundation when attached to the required utilities. Does not include recreational vehicles.

Marsh – Wetland dominated by herbaceous or non-woody plants often developing in shallow ponds or depressions, river margins, tidal areas, and estuaries.

Masonry – Built-up construction of building units made of clay, shale, concrete, glass, gypsum, stone, or other approved units bonded together with or without mortar or grout or other accepted methods of joining.

Mean return period – The average time (in years) between landfall or nearby passage of a tropical storm or hurricane.

Mean water elevation – The surface across which waves propagate. The mean water elevation is calculated as the stillwater elevation plus the wave setup.

Mean sea level (MSL) – Average height of the sea for all stages of the tide, usually determined from hourly height observations over a 19-year period on an open coast or in adjacent waters having free access to the sea. See also *National Geodetic Vertical Datum*.

Metal roof panel – Interlocking metal sheet having a minimum installed weather exposure of 3 square feet per sheet.

Minimal Wave Action area (MiWA) – The portion of the coastal Special Flood Hazard Area where base flood wave heights are less than 1.5 feet.

Mitigation – Any action taken to reduce or permanently eliminate the long-term risk to life and property from natural hazards.

Mitigation Directorate – Component of the Federal Emergency Management Agency directly responsible for administering the flood hazard identification and floodplain management aspects of the National Flood Insurance Program.

Moderate Wave Action area (MoWA) – See *Coastal A Zone.*

National Flood Insurance Program (NFIP) – Federal program created by Congress in 1968 that makes flood insurance available in communities that enact and enforce satisfactory floodplain management regulations.

National Geodetic Vertical Datum (NGVD) – Datum established in 1929 and used as a basis for measuring flood, ground, and structural elevations, previously referred to as Sea Level Datum or Mean Sea Level. The Base Flood Elevations shown on most of the Flood Insurance Rate Maps issued by the Federal Emergency Management Agency are referenced to NGVD or, more recently, to the *North American Vertical Datum.*

Naturally decay-resistant wood – Wood whose composition provides it with some measure of resistance to decay and attack by insects, without preservative treatment (e.g., heartwood of cedar, black locust, black walnut, and redwood).

New construction – *For the purpose of determining flood insurance rates* under the National Flood Insurance Program, structures for which the start of construction commenced on or after the effective date of the initial Flood Insurance Rate Map or after December 31, 1974, whichever is later, including any subsequent improvements to such structures. (See also *Post-FIRM structure.*) *For floodplain management purposes*, new construction means structures for which the start of construction commenced on or after the effective date of a floodplain management regulation adopted by a community and includes any subsequent improvements to such structures.

Non-load-bearing wall – Wall that does not support vertical loads other than its own weight. See also *Load-bearing wall.*

Nor'easter – A type of storm that occurs along the East Coast of the United States where the wind comes from the northeast. Nor'easters can cause coastal flooding, coastal erosion, hurricane-force winds, and heavy snow.

North American Vertical Datum (NAVD) – Datum established in 1988 and used as a basis for measuring flood, ground, and structural elevations. NAVD is used in many recent Flood Insurance Studies rather than the National Geodetic Vertical Datum.

Open foundation – A foundation that allows water to pass through the foundation of an elevated building, which reduces the lateral flood loads the foundation must resist. Examples of open foundations are pile, pier, and column foundations.

Operational costs – Costs associated with the use of the building, such as the cost of utilities and insurance. Optimizing energy efficiency may result in a higher initial cost but save in operational costs.

Oriented strand board (OSB) – Mat-formed wood structural panel product composed of thin rectangular wood strands or wafers arranged in oriented layers and bonded with waterproof adhesive.

Overwash – Occurs when low-lying coastal lands are overtopped and eroded by storm surge and waves such that the eroded sediments are carried landward by floodwaters, burying uplands, roads, and at-grade structures.

Pier foundation – Foundation consisting of isolated masonry or cast-in-place concrete structural elements extending into firm materials. Piers are relatively short in comparison to their width, which is usually greater than or equal to 12 times their vertical dimension. Piers derive their load-carrying capacity through skin friction, end bearing, or a combination of both.

Pile foundation – Foundation consisting of concrete, wood, or steel structural elements driven or jetted into the ground or cast-in-place. Piles are relatively slender in comparison to their length, which usually exceeds 12 times their horizontal dimension. Piles derive their load-carrying capacity through skin friction, end bearing, or a combination of both.

Platform framing – A floor assembly consisting of beams, joists, and a subfloor that creates a platform that supports the exterior and interior walls.

Plywood – Wood structural panel composed of plies of wood veneer arranged in cross-aligned layers. The plies are bonded with an adhesive that cures when heat and pressure are applied.

Post-FIRM structure – For purposes of determining insurance rates under the National Flood Insurance Program, structures for which the start of construction commenced on or after the effective date of an initial Flood Insurance Rate Map or after December 31, 1974, whichever is later, including any subsequent improvements to such structures. This term should not be confused with the term new construction as it is used in floodplain management.

Post foundation – Foundation consisting of vertical support members set in holes and backfilled with compacted material. Posts are usually made of wood and usually must be braced. Posts are also known as columns, but columns are usually made of concrete or masonry.

Precast concrete – Structural concrete element cast elsewhere than its final position in the structure. See also *Cast-in-place concrete.*

Pressure-treated wood – Wood impregnated under pressure with compounds that reduce the susceptibility of the wood to flame spread or to deterioration caused by fungi, insects, or marine borers.

Premium – Amount of insurance coverage.

Primary frontal dune – Under the National Flood Insurance Program, a continuous or nearly continuous mound or ridge of sand with relatively steep seaward and landward slopes immediately landward and adjacent to the beach and subject to erosion and overtopping from high tides and waves during major coastal storms. The inland limit of the primary frontal dune occurs at the point where there is a distinct change from a relatively steep slope to a relatively mild slope.

Rating factor (insurance) – A factor used to determine the amount to be charged for a certain amount of insurance coverage (premium).

Recurrence interval – The frequency of occurrence of a natural hazard as referred to in most design codes and standards.

Reinforced concrete – Structural concrete reinforced with steel bars.

Relocation – The moving of a structure to a location that is less prone to flooding and flood-related hazards such as erosion.

Residual risk – The level of risk that is not offset by hazard-resistant design or insurance, and that must be accepted by the property owner.

Retrofit – Any change or combination of adjustments made to an existing structure intended to reduce or eliminate damage to that structure from flooding, erosion, high winds, earthquakes, or other hazards.

Revetment – Facing of stone, cement, sandbags, or other materials placed on an earthen wall or embankment to protect it from erosion or scour caused by flood waters or wave action.

Riprap – Broken stone, cut stone blocks, or rubble that is placed on slopes to protect them from erosion or scour caused by flood waters or wave action.

Risk – Potential losses associated with a hazard, defined in terms of expected probability and frequency, exposure, and consequences. Risk is associated with three factors: threat, vulnerability, and consequence.

Risk assessment – Process of quantifying the total risk to a coastal building (i.e., the risk associated with all the significant natural hazards that may impact the building).

Risk category – As defined in American Society of Civil Engineers (ASCE) 7-10 and the 2012 International Building Code, a building's risk category is based on the risk to human life, health, and welfare associated with potential damage or failure of the building. These risk categories dictate which design event is used when calculating performance expectations of the building, specifically the loads the building is expected to resist.

Risk reduction – The process of reducing or offsetting risks. Risk reduction is comprised of two aspects: physical risk reduction and risk management through insurance.

Risk tolerance – Some owners are willing and able to assume a high degree of financial and other risks, while other owners are very conservative and seek to minimize potential building damage and future costs.

Riverine SFHA – The portion of the Special Flood Hazard Area mapped as Zone AE and where the source of flooding is riverine, not coastal.

Roof deck – Flat or sloped roof surface not including its supporting members or vertical supports.

Sand dunes – Under the National Flood Insurance Program, natural or artificial ridges or mounds of sand landward of the beach.

Scour – Removal of soil or fill material by the flow of flood waters. Flow moving past a fixed object accelerates, often forming eddies or vortices and scouring loose sediment from the immediate vicinity of the object. The term is frequently used to describe storm-induced, localized conical erosion around pilings and other foundation supports, where the obstruction of flow increases turbulence. See also *Erosion*.

Seawall – Solid barricade built at the water's edge to protect the shore and prevent inland flooding.

Setback – For the purpose of this Manual, a State or local requirement that prohibits new construction and certain improvements and repairs to existing coastal buildings in areas expected to be lost to shoreline retreat.

Shearwall – Load-bearing wall or non-load-bearing wall that transfers in-plane lateral forces from lateral loads acting on a structure to its foundation.

Shoreline retreat – Progressive movement of the shoreline in a landward direction; caused by the composite effect of all storms over decades and centuries and expressed as an annual average erosion rate. Shoreline retreat is essentially the horizontal component of erosion and is relevant to long-term land use decisions and the siting of buildings.

Single-ply membrane – Roofing membrane that is field-applied with one layer of membrane material (either homogeneous or composite) rather than multiple layers. The four primary types of single-ply membranes are chlorosulfonated polyethylene (CSPE) (Hypalon), ethylene propylene diene monomer (EPDM), polyvinyl chloride (PVC), and thermoplastic polyolefin (TPO).

Siting – Choosing the location for the development or redevelopment of a structure.

Special Flood Hazard Area (SFHA) – Under the National Flood Insurance Program, an area having special flood, mudslide (i.e., mudflow), or flood-related erosion hazards, and shown on a Flood Hazard Boundary Map or Flood Insurance Rate Map as Zone A, AO, A1-A30, AE, A99, AH, V, V1-V30, VE, M, or E. The area has a 1 percent chance, or greater, of flooding in any given year.

Start of construction (for other than new construction or substantial improvements under the Coastal Barrier Resources Act) – Under the National Flood Insurance Program, date the building permit was issued, provided the actual start of construction, repair, reconstruction, rehabilitation, addition placement, or other improvement was within 180 days of the permit date. The actual start means either the first placement of permanent construction of a structure on a site such as the pouring of slab or footings, the installation of piles, the construction of columns, or any work beyond the stage of excavation; or the placement of a manufactured home on a foundation. Permanent construction does not include land preparation, such as clearing, grading, and filling; nor the installation of streets or walkways; excavation for a basement, footings, piers, or foundations or the erection of temporary forms; or the installation on the property of accessory buildings, such as garages or sheds not occupied as dwelling units or not part of the main structure. For a substantial improvement, the actual start of construction means the first alteration of any wall, ceiling, floor, or other structural part of a building, whether or not that alteration affects the external dimensions of the building.

State Coordinating Agency – Under the National Flood Insurance Program, the agency of the State government, or other office designated by the Governor of the State or by State statute to assist in the implementation of the National Flood Insurance Program in that State.

Stillwater elevation – The elevations of the water surface resulting solely from storm surge (i.e., the rise in the surface of the ocean due to the action of wind and the drop in atmospheric pressure association with hurricanes and other storms).

Storm surge – Water pushed toward the shore by the force of the winds swirling around a storm. It is the greatest cause of loss of life due to hurricanes.

Storm tide – Combined effect of storm surge, existing astronomical tide conditions, and breaking wave setup.

Structural concrete – All concrete used for structural purposes, including plain concrete and reinforced concrete.

Structural fill – Fill compacted to a specified density to provide structural support or protection to a structure. See also *Fill*.

Structure – *For floodplain management purposes* under the National Flood Insurance Program (NFIP), a walled and roofed building, gas or liquid storage tank, or manufactured home that is principally above ground. *For insurance coverage purposes* under the NFIP, structure means a walled and roofed building, other than a gas or liquid storage tank, that is principally above ground and affixed to a permanent site, as well as a manufactured home on a permanent foundation. For the latter purpose, the term includes a building undergoing construction, alteration, or repair, but does not include building materials or supplies intended for use in such construction, alteration, or repair, unless such materials or supplies are within an enclosed building on the premises.

Substantial damage – Under the National Flood Insurance Program, damage to a building (regardless of the cause) is considered substantial damage if the cost of restoring the building to its before-damage condition would equal or exceed 50 percent of the market value of the structure before the damage occurred.

Substantial improvement – Under the National Flood Insurance Program, improvement of a building (such as reconstruction, rehabilitation, or addition) is considered a substantial improvement if its cost equals or exceeds 50 percent of the market value of the building before the start of construction of the improvement. This term includes structures that have incurred substantial damage, regardless of the actual repair work performed. The term does not, however, include either (1) any project for improvement of a structure to correct existing violations of State or local health, sanitary, or safety code specifications which have been identified by the local code enforcement official and which are the minimum necessary to ensure safe living conditions, or (2) any alteration of a "historic structure," provided that the alteration will not preclude the structure's continued designation as a "historic structure."

Super typhoons – Storms with sustained winds equal to or greater than 150 mph.

Threat – The probability that an even of a given recurrence interval will affect the building within a specified period. See *Risk*.

Tornado – A rapidly rotating vortex or funnel of air extending groundward from a cumulonimbus cloud

Tributary area – The area of the floor, wall, roof, or other surface that is supported by the element. The tributary area is generally a rectangle formed by one-half the distance to the adjacent element in each applicable direction.

Tropical cyclone – A low-pressure system that generally forms in the tropics, and is often accompanied by thunderstorms.

Tropical depression – Tropical cyclone with some rotary circulation at the water surface. With maximum sustained wind speeds of up to 39 miles per hour, it is the second phase in the development of a hurricane.

Tropical disturbance – Tropical cyclone that maintains its identity for at least 24 hours and is marked by moving thunderstorms and with slight or no rotary circulation at the water surface. Winds are not strong. It is a common phenomenon in the tropics and is the first discernable stage in the development of a hurricane.

Tropical storm – Tropical cyclone that has 1-minute sustained wind speeds averaging 39 to 74 miles per hour (mph).

Tsunami – Long-period water waves generated by undersea shallow-focus earthquakes, undersea crustal displacements (subduction of tectonic plates), landslides, or volcanic activity.

Typhoon – Name given to a hurricane in the area of the western Pacific Ocean west of 180 degrees longitude.

Underlayment – One or more layers of felt, sheathing paper, non-bituminous saturated felt, or other approved material over which a steep-sloped roof covering is applied.

Undermining – Process whereby the vertical component of erosion or scour exceeds the depth of the base of a building foundation or the level below which the bearing strength of the foundation is compromised.

Uplift – Hydrostatic pressure caused by water under a building. It can be strong enough lift a building off its foundation, especially when the building is not properly anchored to its foundation.

Variance – Under the National Flood Insurance Program, grant of relief by a community from the terms of a floodplain management regulation.

Violation – Under the National Flood Insurance Program (NFIP), the failure of a structure or other development to be fully compliant with the community's floodplain management regulations. A structure or other development without the elevation certificate, other certifications, or other evidence of compliance required in Sections 60.3(b)(5), (c)(4), (c)(10), (d)(3), (e)(2), (e)(4), or (e)(5) of the NFIP regulations is presumed to be in violation until such time as that documentation is provided.

Vulnerability – Weaknesses in the building or site location that may result in damage. See *Risk*.

Water surface elevation – Under the National Flood Insurance Program, the height, in relation to the National Geodetic Vertical Datum of 1929 (or other datum, where specified), of floods of various magnitudes and frequencies in the floodplains of coastal or riverine areas.

Wave – Ridge, deformation, or undulation of the water surface.

Wave height – Vertical distance between the wave crest and wave trough. Wave crest elevation is the elevation of the crest of a wave, referenced to the National Geodetic Vertical Datum, North American Vertical Datum, or other datum.

Wave overtopping – Occurs when waves run up and over a dune or barrier.

Wave runup – Is the rush of water up a slope or structure. Wave runup occurs as waves break and run up beaches, sloping surfaces, and vertical surfaces.

Wave runup depth – At any point is equal to the maximum wave runup elevation minus the lowest eroded ground elevation at that point.

Wave runup elevation – Is the elevation reached by wave runup, referenced to the National Geodetic Vertical Datum or other datum.

Wave setup – Increase in the stillwater surface near the shoreline due to the presence of breaking waves. Wave setup typically adds 1.5 to 2.5 feet to the 100-year stillwater flood elevation and should be discussed in the Flood Insurance Study.

Wave slam – The action of wave crests striking the elevated portion of a structure.

Wet floodproofing – A flood retrofitting technique that involves modifying a structure to allow floodwaters to enter it in such a way that damage to a structure and its contents is minimized.

Zone A – Under the National Flood Insurance Program, area subject to inundation by the 100-year flood where wave action does not occur or where waves are less than 3 feet high, designated Zone A, AE, A1-A30, A0, AH, or AR on a Flood Insurance Rate Map.

Zone AE – The portion of the Special Flood Hazard Area (SFHA) not mapped as Zone VE. It includes the Moderate Wave Action area, the Minimal Wave Action area, and the riverine SFHA.

Zone B – Areas subject to inundation by the flood that has a 0.2-percent chance of being equaled or exceeded during any given year, often referred to the as 500-year flood. Zone B is provided on older flood maps, on newer maps this is referred to as "shaded Zone X."

Zone C – Designates areas where the annual probability of flooding is less than 0.2 percent. Zone C is provided on older flood maps, on newer maps this is referred to as "unshaded Zone X."

Zone V – See *Coastal High Hazard Area*.

Zone VE – The portion of the coastal Special Flood Hazard Area where base flood wave heights are 3 feet or greater, or where other damaging base flood wave effects have been identified, or where the primary frontal dune has been identified.

Zone X – Under the National Flood Insurance Program, areas where the flood hazard is lower than that in the Special Flood Hazard Area. Shaded Zone X shown on recent Flood Insurance Rate Maps (Zone B on older maps) designate areas subject to inundation by the 500-year flood. Unshaded Zone X (Zone C on older Flood Insurance Rate Maps) designate areas where the annual probability of flooding is less than 0.2 percent.

Zone X (Shaded) – Areas subject to inundation by the flood that has a 0.2-percent chance of being equaled or exceeded during any given year, often referred to the as 500-year flood.

Zone X (Unshaded) – Designates areas where the annual probability of flooding is less than 0.2 percent.

Index, Volume I

Bold text indicates chapter titles or major headings. Italicized text indicates a figure or table.